W9-BSC-648

Sonochemistry

Timothy J. Mason

Professor of Chemistry, University of Coventry

Series sponsor: **ZENECA**

ZENECA is a major international company active in four main areas of business: Pharmaceuticals, Agrochemicals and Seeds, Specialty Chemicals, and Biological Products.

ZENECA's skill and innovative ideas in organic chemistry and bioscience create products and services which improve the world's health, nutrition, environment, and quality of life.

ZENECA is committed to the support of education in chemistry and chemical engineering.

OXFORD
UNIVERSITY PRESS

OXFORD
UNIVERSITY PRESS

Great Clarendon Street, Oxford OX2 6DP

Oxford University Press is a department of the University of Oxford
and furthers the University's aim of excellence in research, scholarship,
and education by publishing worldwide in

Oxford New York

Athens Auckland Bangkok Bogotá Buenos Aires Calcutta
Cape Town Chennai Dar es Salaam Delhi Florence Hong Kong Istanbul
Karachi Kuala Lumpur Madrid Melbourne Mexico City Mumbai
Nairobi Paris São Paulo Singapore Taipei Tokyo Toronto Warsaw

and associated companies in Berlin Ibadan

Oxford is a registered trade mark of Oxford University Press

Published in the United States
by Oxford University Press Inc., New York

First published 1999

British Library Cataloguing in Publication Data
Data available

Library of Congress Cataloging in Publication Data
(Data applied for)

ISBN 0 19 850371 7 (Pbk)

Typeset by EXPO Holdings, Malaysia

Printed in Great Britain
on acid-free paper by Bath Press, Bath,

Series Editor's Foreword

Oxford Chemistry Primers are designed to provide clear and concise introductions to a wide range of topics that may be encountered by chemistry students as they progress from the freshman stage through to graduation. The Physical Chemistry series will contain books easily recognised as relating to established fundamental core material that all chemists will need to know, as well as books reflecting new directions and research trends in the subject, thereby anticipating (and perhaps encouraging) the evolution of modern undergraduate courses.

In this Physical Chemistry Primer Professor Tim Mason presents a stimulating self-contained and practically oriented introduction to the principles and applications of *Sonochemistry*. The Primer will interest all students (and their mentors) who require an easy-to-read yet authoritative introduction to this increasingly important topic.

Richard G. Compton
Physical and Theoretical Chemistry Laboratory,
University of Oxford

Preface

It has been recognised for many years that power ultrasound has great potential for use in a wide variety of processes in the chemical and allied industries. Reported applications include cleaning, sterilisation, flotation, drying, degassing, defoaming, soldering, plastic welding, drilling, filtration, homogenisation, emulsification, dissolution, deaggregàtion of powders, biological cell disruption, extraction, crystallisation, and, more recently, as a stimulus for chemical reactions.

With the increasing exploitation of power ultrasound in chemistry and its inclusion in university courses comes the need for specialist text books covering this discipline. Several advanced texts exist but there is currently nothing in print which offers a basic introduction to the subject for undergraduate readers. In 1991 I wrote a text entitled "Practical Sonochemistry, A users guide to applications in chemistry and chemical engineering" which covered the basic aspects of theory and laboratory practise but is now out of print. This primer updates and extends the information which was previously published in that text. It is designed to answer the following questions:

What is sonochemistry?
Why is it important?
What systems are affected by sonochemistry?
What are the major applications of this technique?
What types of equipment are available and how do they compare?
How is the equipment best configured in the laboratory?
What parameters can be changed to optimise sonochemical results?
What equipment is available for the scale-up of sonochemistry?

An ever increasing number of chemists are using power ultrasound to promote synthetic reactions yet many experimentalists experience difficulty in producing significant sonochemical effects. On the whole this problem appears to be the result of a lack of appreciation of the correct methodologies to adopt in order to introduce power ultrasound into a reacting medium. This in turn is almost certainly because of a lack of understanding of the principles of ultrasound and acoustic cavitation—subjects which are not very common in the background of practising chemists. In Chapter 1 you will find a general introduction to sonochemistry. This includes experimental methods which have been used for the determination of the cavitation threshold for a medium, i.e. the minimum ultrasonic power level at which cavitation can be induced and sonochemistry becomes possible. This chapter also reviews the methods available for the measurement of the ultrasonic power level entering a reaction. Chapters 2 and 3 contain practical details on the construction and method of use of ultrasonic baths and probe systems, respectively, while Chapter 4 explores the type of equipment which is currently available for large-scale sonochemistry. I hope that these four chapters contain the basic information necessary for the chemist to understand how to use power ultrasound in reactions, and for the engineer to appreciate some of the important design implications when introducing sonochemistry or ultrasonic processing on a plant scale.

This book is suitable for chemists and chemical engineers at all levels who wish to gain a rapid insight into what is becoming a standard method in the chemical laboratory.

Contents

I would like to dedicate this book to my family for suffering through another summer of book writing, and in particular to my wife Christine and son Daniel for their extended and repeated sessions of proof reading.

1 An introduction to the uses of power ultrasound in chemistry

If you were asked what you knew about ultrasound you would almost certainly start with the fact that it is used in animal communications (e.g. bat navigation and dog whistles). You might then recall that ultrasound is used in medicine for fetal imaging, in underwater range finding (SONAR), or in the non-destructive testing of materials for flaws. For a chemist, however, sound would probably not be the first form of energy that would be considered for the stimulation of a chemical reaction. Nowadays an increasing number of scientists are becoming interested in a new field of research—sonochemistry—a term primarily used to describe the effect of ultrasonic sound waves on chemical reactions, but also applied to processing involving power ultrasound. The name is derived from the prefix *Sono* indicating sound, paralleling the longer established techniques which use light (*photo*chemistry) and electricity (*electro*chemistry) to achieve chemical activation. However unlike most other chemical technologies which require some special attribute of the system in order to function, e.g. the use of microwaves (dipolar species), electrochemistry (conducting medium), and photochemistry (the presence of a chromophore, i.e. a grouping capable of activation by light irradiation), ultrasound requires only the presence of a liquid to transmit its power. In this sense sonochemistry can be considered to be a general activation technique like thermochemistry (heat) and piezochemistry (pressure).

The first commercial application of ultrasonics dates back to 1917 with the echo sounding technique of Langevin for the estimation of the depth of water. The discovery was the direct result of an idea which arose from suggestions generated by a competition organised in 1912 to find a non-visual method of detecting icebergs in the open sea and so avoid any repetition of the disaster which befell the Titanic. The early 'echo sounder' simply sent a pulse of ultrasound from the keel of a boat to the bottom of the sea from which it was reflected back to a detector also on the keel. For sound waves, since the distance travelled through a medium = $\frac{1}{2} \times$ time $\times$ velocity (and the velocity of sound in seawater is accurately known), the distance to the bottom could be gauged from the time taken for the signal to return to the boat. If some foreign object (e.g. a submarine) were to come between the boat and the bottom of the seabed an echo would be produced from this in advance of the bottom echo. This system was very important to the Allied Submarine

Detection Investigation Committee during the Second World War and became popularly known by the acronym ASDIC. Later developments resulted in the system known as SONAR (SOund Navigation And Ranging) which allowed the surrounding sea to be scanned. As an example of its efficiency, using modern SONAR it is possible to locate a small fish only 35 cm in length at a depth of 500 metres. It is interesting to note that the original ASDIC system actually came before the corresponding RAdio Detection And Ranging System (RADAR) by 30 years.

Essentially all imaging from medical ultrasound to non-destructive testing relies upon the same pulse–echo type of approach but with considerably refined electronic hardware. The refinements enable the equipment not only to detect reflections of the sound wave from the hard, metallic surface of a submarine in water but also much more subtle changes in the media through which sound passes (e.g. those between different tissue structures in the body). It is high frequency ultrasound (in the range 2 to 10 MHz) which is used primarily in this type of application because by using these much shorter wavelengths it is possible to detect much smaller areas of phase change, i.e. give better 'definition'.

1.1 The power of sound

Sound, as a general subject for study, is traditionally found in a physics syllabus, but it is not a topic which is met in a chemistry course and so is somewhat unfamiliar to practising chemists. Sound is transmitted through a medium by inducing vibrational motion of the molecules through which it is travelling. This motion can be visualised as rather like the ripples produced when a pebble is dropped into a pool of still water. The waves move but the water molecules which constitute the wave revert to their normal positions after the wave has passed. An alternative representation is provided by the effect of a sudden twitch of the end of a horizontal stretched spring. Here the vibrational energy is transmitted through the spring as a compression wave which is seen to traverse its whole length. This is just a single compression wave and it does not equate to sound itself which is a whole series of such compression waves separated by rarefaction (stretching) waves in between. The pitch (or note) of the sound produced by this series of waves depends upon their frequency, i.e. the number of waves which pass a fixed point in unit time. For middle C this is 256 per second. Sound waves can be represented as a series of vertical lines or shaded colour where line separation or colour depth represent intensity, or as a sine wave (Fig. 1.1). Here P_A is the ambient pressure in the fluid and the sine wave represents pressure variation with position at a fixed point in time. The wave amplitude is P_W and the wavelength is λ.

The physical effects of sound vibrations are most easily experienced by standing in front of a loudspeaker playing music at high volume. The actual sound vibrations are transmitted through the air and are not only audible but

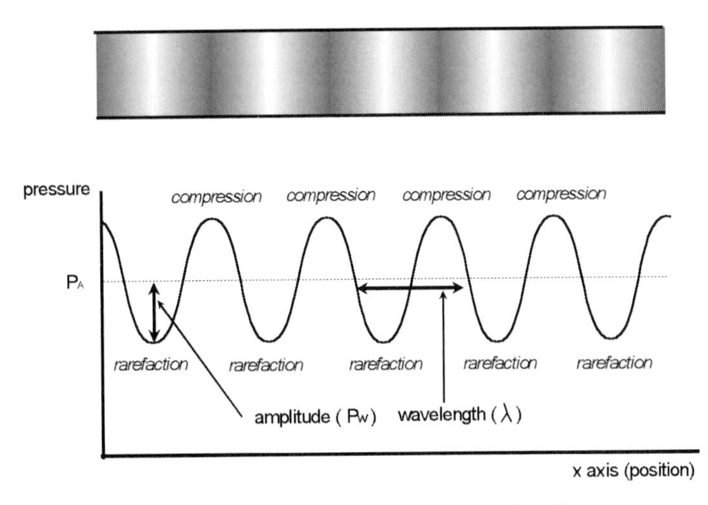

Figure 1.1 Representations of sound motion

can also be sensed by the body through the skin. The bass notes are felt through the body more easily than the high notes and this is connected with the frequency of the pressure pulse creating the sound. Low frequency sound becomes audible at around 18 Hz (1 Hz = 1 Hertz = 1 cycle per second), but as the frequency of the sound is raised (becoming more treble) it becomes more difficult for the body to respond and that sensation is lost. High frequency sound, while not noticeably affecting the body, does cause severe annoyance to hearing, e.g. feedback noise from a microphone through a loud speaker. At even higher frequencies the ear finds it difficult to respond and eventually the human hearing threshold is reached, normally around 18–20 kHz for adults, sound beyond this limit is inaudible and is defined as ultrasound. The hearing threshold is not the same for other animal species, thus dogs respond to ultrasonic whistles (so called 'silent' dog whistles) and bats use frequencies well above 50 kHz for navigation (Fig. 1.2).

The broad classification of ultrasound as sound above 20 kHz and up to 100 MHz can be subdivided into two distinct regions Power and Diagnostic. The former is generally at lower frequency where greater acoustic energy can be generated to induce cavitation in liquids. It is cavitation which is the origin of sonochemical effects. Sonochemistry normally uses frequencies between 20 and 40 kHz simply because this is the range employed in common laboratory equipment. However since acoustic cavitation in liquids can be generated well above these frequencies, recent researchers into sonochemistry use a much broader range (Fig. 1.2). High frequency ultrasound from around 5 MHz and above does not produce cavitation and this is the frequency range used in medical imaging.

A whistle which generates a frequency of 20 kHz is inaudible to humans but perfectly audible to a dog—and produces no physical harm to either. It is

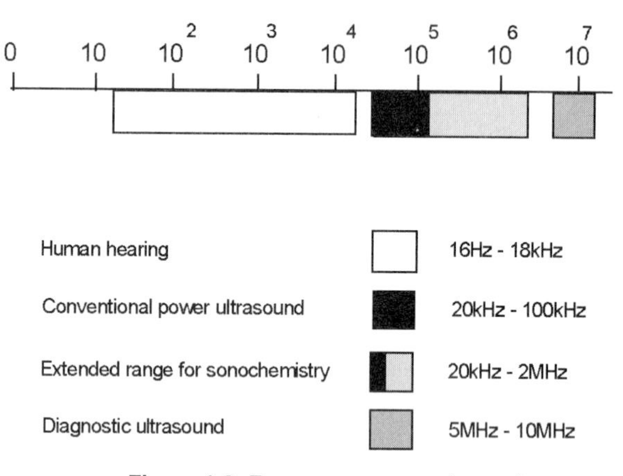

Figure 1.2 Frequency ranges of sound

however in the correct FREQUENCY range to affect chemical reactivity (Power Ultrasound). Yet such a whistle blown in a laboratory will not influence chemical reactions in any way. This is because the whistle is producing sound energy in air and airborne sound cannot be transferred into a liquid. Technically this is due to an impedance mismatch between the two. Different materials have different 'resistances' to the passage of sound which are determined by their elastic properties and cross-sectional areas. Efficient energy transfer between two materials is only possible when the resistances are balanced.

The topic of diagnostic ultrasound is not within the scope of this text but a number of aspects of its use in food technology—quite close to materials processing—are to be found elsewhere [1]. A summary of the uses of diagnostic ultrasound in chemistry appears in Table 1.1.

Table 1.1 Some uses of diagnostic ultrasound in chemistry

Type of diagnostic ultrasound	Use of this technique
Pulse echo (echo ranging)	To estimate the volume of material remaining in a vat or reagent holder (accurate measurement of distance from an interface)
Sound velocity measurement	To monitor the progress of a reaction (change from starting material to product is reflected in a change in sound velocity through a medium)
Sound attenuation measurement	To monitor composition of a product in quality control (a steady attenuation indicates constant composition)

1.2 The importance of power ultrasound in industry

It has been recognised for many years that power ultrasound has great potential for use in a wide variety of processes in the chemical and allied industries. Some of these applications have been known and used for many years [2,3,4] while others are undergoing a renaissance and developing into new and exciting possibilities, as in the use of power ultrasound in therapeutic medicine (Table 1.2). Two of these applications have provided the direct antecedents of the types of equipment now commonly used for sonochemistry, namely ultrasonic welders and cleaning baths.

Table 1.2 Some industrial uses of power ultrasound

Field	Application
Plastic welding	Fabrication of thermoplastic articles
Cleaning	Cleaning in aqueous media of engineering items, medical instruments, and jewellery
Cutting	Accurate drilling and cutting of all forms of material from ceramics to food products
Therapeutic medicine	Dissolution of blood clots, enhanced chemotherapy
Processing	Pigments and solid dispersion in liquid media, crystallisation, filtration, drying, degassing, defoaming, homogenisation, emulsification, dissolution, deaggregation, extraction
Sonochemistry	Electrochemistry, environmental protection, catalysis, benign synthesis

Ultrasonic welding

A large proportion of ultrasonic equipment currently in industry is involved in welding or rivetting plastic mouldings for the consumer market. This equipment consists of a generator producing an alternating electrical potential at a frequency of around 20 kHz which feeds a transducer (normally piezoelectric or magnetostrictive, see below), which converts the electrical energy into mechanical energy. A shaped tool or horn transmits (and amplifies) the vibrating motion to a shaped die pressing together the two pieces of material to be welded. The vibrational amplitude is typically 50–100 microns.

Ultrasonic welding is generally used for the more rigid amorphous types of thermoplastic. Thermoplastics have two properties which make them particularly suited to ultrasonic welding: (a) low thermal conductivity and (b) melting or softening temperatures of between 100 and 200 °C. Ultrasonic vibrations pass through the bulk plastic of the component, without producing bulk heating, to the joint which rapidly heats up. As soon as the ultrasonic power is switched off the bulk material becomes a heat-sink, cooling the welded joint. This is a great improvement on conductive heating for welding since in this case the thermal gradient has to be reversed before cooling occurs. This leads to long heating/cooling process cycles and possible distortion of the material. Another major advantage of the use of ultrasound is the high joint strength of the weld, reaching 90–98 per cent of the material strength. Indeed test samples usually break in the body of the material and not at the weld itself.

Ultrasonic cleaning

Ultrasonic cleaning is another major application for power ultrasound. It is now such a well established technique that laboratories without access to an ultrasonic cleaning bath are in a minority. It is important to recognise the historical significance of the development of ultrasonic cleaning bath technology on the growth of sonochemistry, because the use of ultrasonic cleaners is probably the first method to which the chemist will turn when starting sonochemistry research.

Although the laboratory ultrasonic cleaning bath is familiar to the chemist, the industrial applications of such cleaning are perhaps less well known. Yet it is developments in industrial cleaning which have made it possible to consider large-scale chemical reactions, since the larger the batch chemical process the bigger will be the size requirement of the bath. Ultrasonic cleaning can be both delicately applied (computer microcomponents, jewellery, medical instruments) and used for very large items (gas cylinder heads, engine blocks).

Cutting

Ultrasonic cutting has been available to industry since the early 1950s specifically for accurate profile cutting of brittle materials such as ceramics and glass. It has been extensively used in the aerospace industry since the 1970s for glass and carbon fibre composites. In recent years, ultrasonic cutting has been introduced into the food processing industry where it would appear to have far wider application than the laser and water (oil) jet cutting technologies introduced in the 1980s [1].

Ultrasonic cutting uses a knife type blade attached through a shaft to an ultrasonic source. Essentially the shaft with its blade behaves as an ultrasonic horn driven normally at 20 kHz and with a generator similar to that of an ultrasonic welder. The cutting action is a combination of the pressure applied to a sharp cutting edge surface and the mechanical longitudinal vibration of the blade. Typically the tip movement will be in the range 50 to 100 microns peak to peak. Several advantages arise from the mode of action of this technology.

1. In conventional cutting the blade has to compress the bulk material to allow a gap the width of the blade to pass through and this applies a tensile rupturing force at the crack tip. With ultrasonic cutting the whole blade moves or vibrates continuously as it stretches and contracts. This very high frequency movement effectively reduces the coefficient of friction to a very low level, enabling the blade to slide more easily through the bulk material.

2. The ultrasonic vibration of 20 kHz applies an oscillating force to the material to be cut and generates a crack (cut) at the tip, controlling its propagation or growth, thereby minimising the stress on the bulk material.

3. The repeated application of the cutting tip to the product applies a local fatiguing effect which reduces significantly the overall force required to break the bonds of the bulk material.

Therapeutic ultrasound

The field of imaging is well accepted in the medical field but in recent years ultrasound has been used in therapeutic medicine (Fig. 1.3) [5]. One such development is for the destruction of blood clots, where a miniaturised ultrasonic device (1 MHz) is attached to the end of a catheter so that it can be inserted into the blood vessel. The device is brought close to the clot and a fibrolytic enzyme is released. When the transducer is activated the chemical, assisted by acoustic energy, rapidly dissolves the clot. Another use involves a focused array of transducers for use in cancer therapy. The array can accurately target small rugby football shaped volumes within the body to thermally destroy cancerous cells in a sequence of exposures which cover the infected region. At lower powers the focused ultrasound can be used to enhance the action of chemotherapy agents, often a porphyrin, which become localised in the region of cancerous cells. In this case the effects are generally thought to enhance the chemical reactions involved in chemotherapy, which are generally considered to be radical in nature.

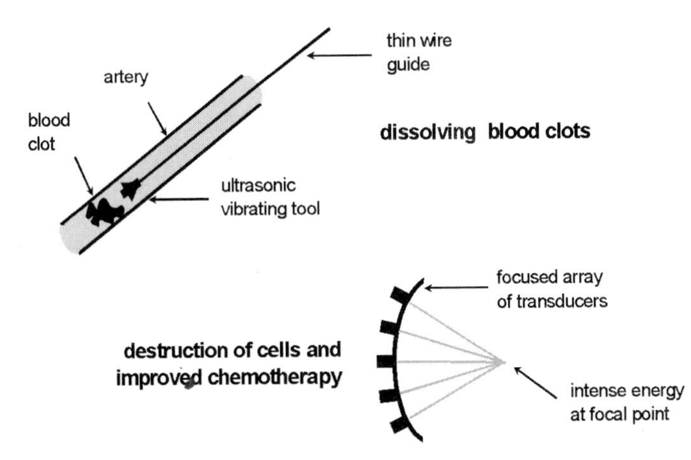

Figure 1.3 Therapeutic applications of ultrasound

Processing

The use of power ultrasound in the processing industry has a long history with a great number of different applications. Some of these are dealt with later in this chapter (Table 1.5).

Sonochemistry

An ever increasing interest in the specific uses of power ultrasound to affect organic, inorganic, and organometallic synthesis (sonochemistry) has resulted in the organisation of several major conferences and the establishment of special interest groups. It was in 1986 that the first ever International Symposium on Sonochemistry was held at Warwick University, UK as part of the Autumn Meeting of the Royal Society of Chemistry [6]. This meeting was significant in that it was the beginning of serious interest in the uses of ultrasound in chemistry as a study in itself; since then there have been regular international meetings on the subject. The formation of a Royal Society of Chemistry Sonochemistry Group in 1987 followed by a European Society in 1990 and then other national groups has meant that the subject has expanded greatly over the last few years.

Uses for ultrasound have been found in chemistry which are quite distinct from synthesis. Amongst such areas are environmental protection (both biological and chemical); material science (new catalytic materials, improved extraction, crystallisation, and new methods in polymer technology); electrochemistry (providing improvements in plating electrosynthesis and electroanalysis); and biotechnology (the modification of enzyme and whole-cell activities).

1.3 Cavitation—the origin of sonochemical effects

The effects of ultrasound on chemical transformations are not the result of any direct coupling of the sound field with the chemical species involved on a molecular level. The reason why power ultrasound is able to produce chemical effects is through the phenomenon of cavitation. Cavitation is the production of microbubbles in a liquid when a large negative pressure is applied to it. This was first characterised at the turn of the century by Sir John Thornycroft and Sidney Barnaby [7]. They were called in to investigate the poor performance of a new screw driven destroyer, HMS *Daring*, the ship did not reach the high speeds expected. The problem was traced to an incorrect setting of the propeller blades resulting in inefficient thrust. As a result of this study, it was found that during the rapid motion of the propeller blade through the water one face produced thrust, as expected, but the trailing edge produced sufficient 'negative' pressure in the water to pull molecules apart and create tiny microbubbles (cavities). These bubbles subsequently collapsed with the release of intense local energy. Cavitational collapse of this type near the metal surface is extremely detrimental to propeller wear since it causes erosion of the blade.

The problem of cavitation caused by propeller motion became a significant problem at the beginning of the Second World War. The Allies had a system of sonar submarine detection which was 'active', that is to say that they sent out a short burst of sound (a ping) and waited to pick up any echo from a submarine. From the time involved between emission and reception of the echo a distance for the target could be estimated. Of course the ping was also detectable by the target which then knew it was under surveillance. The German forces had developed a 'passive' sonar system which simply listened for any underwater noises generated by submarines. That this noise was the result of cavitation produced by the propellers was proven by the fact that the cavitation noise was easily detectable when the submarine was running at periscope depth (some 50 feet under the surface), but became less and less as the ship dived. As we will see later this can be explained by the increase in water pressure with depth which makes cavitation more difficult to produce. During this time there was an enormous amount of research involving sonar which did much to progress the development of transducers. Today many naval research laboratories are interested in the production of harder alloys for propellers and better design to help combat cavitational damage.

How then can power ultrasound produce cavitation? Like any sound wave, ultrasound is transmitted via waves which alternately compress and stretch the molecular spacing of the medium through which it passes (see Fig. 1.1). Thus the average distance between the molecules in a liquid will vary as the molecules oscillate about their mean position. If a large negative pressure, i.e. sufficiently below ambient, is applied to the liquid (here it is the acoustic pressure on rarefaction) so that the distance between the molecules exceeds the critical molecular distance necessary to hold the liquid intact, the liquid will break down and voids will be created, i.e. cavitation bubbles will form. Theoretical calculations indicate that for pure water the negative pressure required is of the order of 10 000 atmospheres. If the calculation is modified so that allowance is made for the bubbles to be filled with vapour (from evaporation of the liquid), cavitation will still require negative pressures of about 1000 atmospheres. In practice cavitation can be produced at considerably lower applied acoustic pressures due to the presence of weak spots in the liquid which lower its tensile strength. Weak spots include the presence of gas nuclei in the form of dissolved gas, minute suspended gas bubbles, or tiny suspended particles. In order to study the true cavitation threshold of a liquid it is necessary to use a degassed liquid which has been subject of ultrafiltration.

When produced in a sound field at sufficiently high power the formation of cavitation bubbles will be initiated during the rarefaction cycle. These bubbles will grow over a few cycles taking in some vapour or gas from the medium (rectified diffusion) to an equilibrium size which matches the frequency of bubble resonance to that of the sound frequency applied. The acoustic field experienced by an individual bubble is not stable because of the interference of other bubbles forming and resonating around it. As a result some bubbles suffer sudden expansion to an unstable size and collapse violently. It is the fate of these cavities when they collapse which generates the energy for chemical and mechanical effects (Fig. 1.4).

There are several theories which have been advanced to explain the energy release involved with cavitation of which the most understandable in a qualitative sense is the 'hot spot' approach [8]. Each cavitation bubble acts as

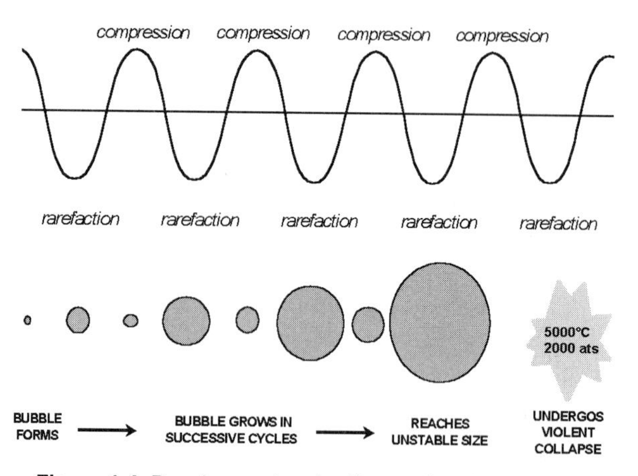

Figure 1.4 Development and collapse of cavitation bubbles

a localised microreactor which, in aqueous systems, generates instantaneous temperatures of several thousand degrees and pressures in excess of one thousand atmospheres.

Over the years it has been common practice to classify cavitation into two types—transient and stable (also called inertial and non-inertial). Transient cavities generally exist for no more than a few acoustic cycles during which time they expand to at least double their initial radius before collapsing violently within a few microseconds. Stable cavities are those which oscillate, often non-linearly, about some equilibrium size with a lifetime of tens of cycles. This is a simplification of the actual situation within an acoustic cavitation zone, but it does emphasise that there is more than one type of cavitation process involved [9].

It was once thought that the spectacular effects attributable to cavitation were entirely due to the collapse of transient cavities. It is now believed that the oscillation of 'stable' bubbles makes a significant contribution to the overall sonochemical effect. In this text, however, we are interested mainly in the overall effects of cavitation and so we will not overly concern ourselves with any distinction between types of cavitation.

1.4 Parameters which affect sonochemistry

In spite of the great amount of published literature in the field of sonochemistry, only a relatively small number of papers report the influence of the reaction conditions on the rate and/or yields of chemical reactions. This is somewhat surprising since it is well known that external parameters have a great influence on cavitation, and since cavitation is necessary to induce sonochemical reactions then it is important that those factors which influence cavitation are understood. Armed with this knowledge it should be easier to organise the experimental conditions such that the sonochemical effects are maximised.

Frequency

As the frequency of irradiation is increased so the rarefaction phase shortens and it is necessary to increase the amplitude (power) of irradiation to maintain an equivalent amount of cavitational energy in the system. In other words more power is required at a higher frequency if the same cavitational effects are to be maintained, e.g. ten times more power is required to make water cavitate at 400 kHz than at 10 kHz.

When the ultrasonic frequency is increased into the MHz region it becomes more and more difficult to produce cavitation in liquids. The simplest explanation for this, is qualitative terms, is that at very high frequency the rarefaction (and compression) cycle is extremely short. The production of a cavity in the liquid requires a finite time to permit the molecules to be pulled apart, so when the rarefaction cycle approaches and becomes shorter than this time, cavitation becomes difficult and then impossible to achieve. It should also be recognised that transducers which operate at these high frequencies are not mechanically capable of generating very high ultrasonic power. This is the main reason why the frequencies generally chosen for cleaning, plastic welding, and subsequently sonochemistry are between 20 and 40 kHz.

Solvent viscosity

The formation and collapse of voids or vapour filled microbubbles (cavities) produces shear forces in the bulk liquid. Since viscosity is a measure of resistance to shear it is more difficult to produce cavitation in a viscous liquid.

Solvent surface tension

Cavitation requires the generation of a liquid–gas interface. Thus it might be expected that employing a solvent of low surface energy per unit area would lead to a reduction in the cavitation threshold. This is not a simple relationship, but certainly where aqueous solutions are involved the addition of a surfactant facilitates cavitation.

Solvent vapour pressure

Cavitation bubbles do not enclose a vacuum. During the expansion phase of cavitation bubble generation, vapour from the surrounding liquid will permeate the interface. This produces a small pressure within the bubble, reducing the pressure differential between cavity and bulk. It is difficult to induce cavitation in a solvent of low vapour pressure because less vapour will enter the bubble. A more volatile solvent will support cavitation at lower acoustic energy and produce vapour filled bubbles. Unfortunately, however, sonochemical effects are based upon the energy produced by cavitation bubble collapse which is cushioned by vapour in the bubble. Hence solvents with high vapour pressures easily generate vapour filled bubbles, but their collapse is cushioned and therefore less energetic.

Bubbled gas

We have already referred to the fact that dissolved gas or small gas bubbles in a fluid can act as nuclei for cavitation. We will also see later that ultrasound can be used to degas a liquid. Thus at the beginning of the sonication of a liquid, any gas which is normally entrapped or dissolved in the liquid promotes cavitation and is removed. Manufacturers of ultrasonic cleaning baths will always wait until the water in their bath is ultrasonically degassed before using it for cleaning, because the bath is not producing its optimum cavitational effects until the gas is removed.

Many research groups deliberately introduce a gas into a sonochemical reaction in order to maintain uniform cavitation. According to theory, the energy developed on collapse of these gas-filled bubbles will be greatest for gases with the largest ratio of specific heats (polytropic index). For this reason monatomic gases (He, Ar, Ne) are used in preference to diatomics (N_2, air, O_2). Gases such as CO_2 are most unsuitable.

External (applied) pressure

Increasing the external pressure will mean that a greater rarefaction pressure is required to initiate cavitation (see Section 1.3 for the effect of depth on cavitation produced by a submarine propeller). More importantly, raising the external pressure will give rise to a larger intensity of cavitational collapse and consequently an enhanced sonochemical effect. Cum *et al.* have studied the oxidation of indane to indan-1-one using potassium permanganate (Scheme 1.1), at two different frequencies (21.5 and 48 kHz) and under various external pressures [10]. For each set of experiments at a particular frequency only the applied pressure was changed while the other parameters (e.g. power, temperature, reaction volume) were kept constant (Fig. 1.5). The results showed that at a specific frequency there is a particular external pressure which will provide an optimum sonochemical reaction, moreover this optimum power depends upon the frequency used.

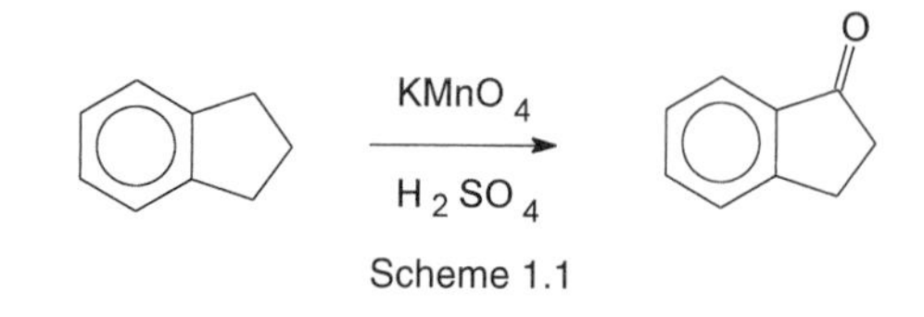

Scheme 1.1

Temperature

Luche and co-workers have extensively studied the Barbier reaction (Scheme 1.2). They have shown that the reaction rate strongly depends on temperature and there is a clearly defined optimum value (Fig. 1.6) [11].

Any increase in temperature will raise the vapour pressure of a medium and so lead to easier cavitation but less violent collapse (see above). This will be accompanied by a decrease in viscosity and surface tension. However,

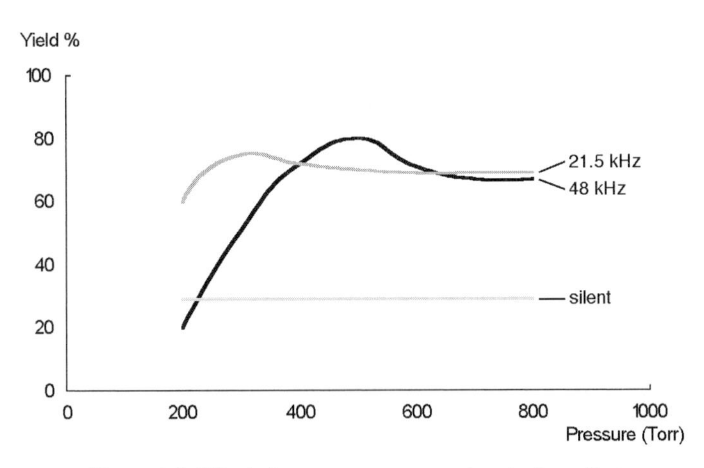

Figure 1.5 Effect of pressure on a sonochemical reaction

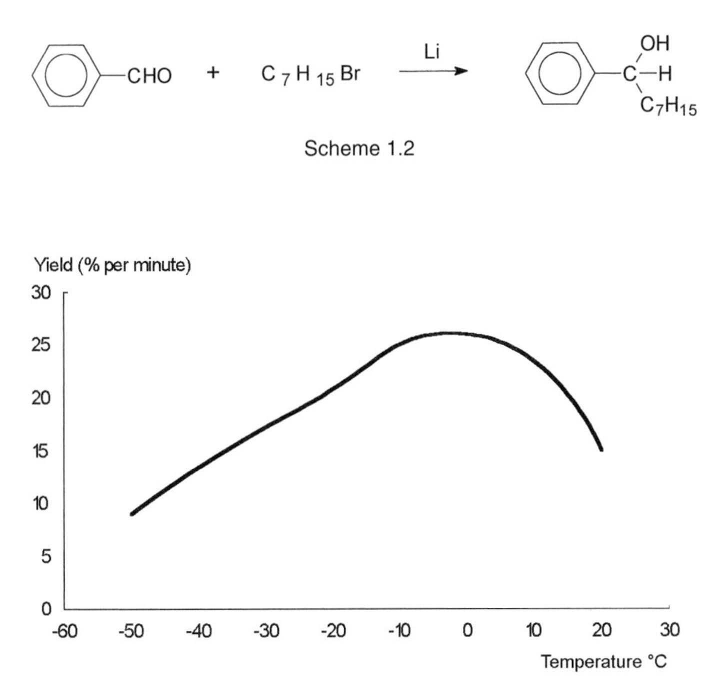

Scheme 1.2

Figure 1.6 Effect of temperature on a sonochemical reaction

at higher temperatures approaching solvent boiling point, a large numbers of cavitation bubbles are generated concurrently. These will act as a barrier to sound transmission and dampen the effective ultrasonic energy from the source which enters the liquid medium.

Intensity

The same reaction was also studied under varying power input (Fig. 1.7) [11]. Once again an optimum value was observed. Power variation was achieved by varying the applied potential (V) at the piezoelectric transducer.

The intensity of sonication is directly proportional to the square of the amplitude of vibration of the ultrasonic source. In general, an increase in intensity will provide an increase in the sonochemical effects, but there are limits to the ultrasonic energy input to the system.

1. A minimum intensity for sonication is required to reach the cavitation threshold. This minimum depends upon the frequency.

2. When a large amount of ultrasonic power enters a system, a great number of cavitation bubbles are generated in the solution. Many of these will coalesce forming larger, more longer lived bubbles. These will certainly act as a barrier to the transfer of acoustic energy through the liquid.

3. At high vibrational amplitude the source of ultrasound will not be able to maintain contact with the liquid throughout the complete cycle. Technically this is known as decoupling, and results in a great loss in efficiency of transfer of power from the source to the medium. Decoupling is more pronounced when large numbers of cavitation bubbles build up at or near the emitting surface of the transducer.

4. The transducer material used in the sonicator will eventually break down as the increasing dimensional changes in the transducer become large enough to fracture the material.

Several examples exist of situations where above a certain energy input the sonochemical effect is reduced, but a particularly good example is the effect of increasing power on the yield of iodine from the sonication of aqueous KI

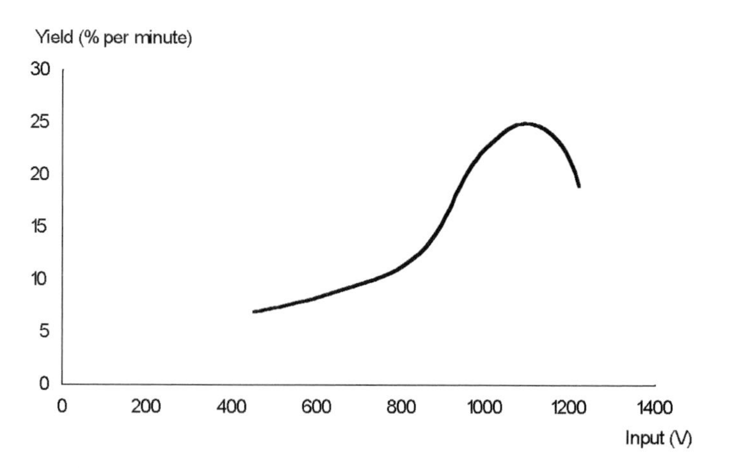

Figure 1.7 Effect of acoustic power on a sonochemical reaction

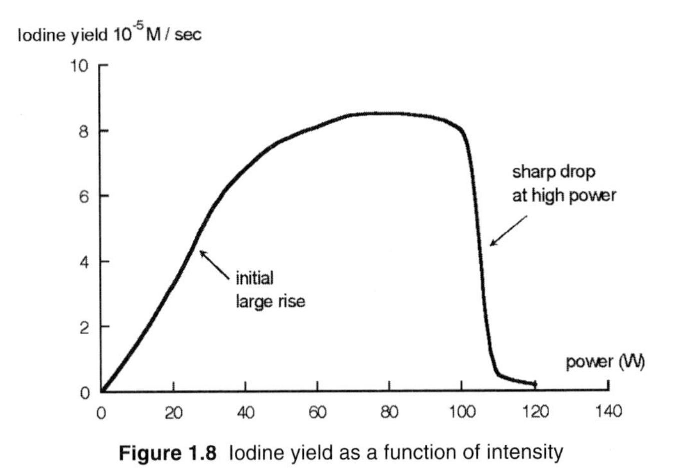

Figure 1.8 Iodine yield as a function of intensity

(Fig. 1.8) [12]. The initial response of iodine yield appears to be proportional to power, but this effect is reduced beyond 40 W and drops dramatically above 100 W where decoupling occurs.

Attenuation of sound

For a variety of reasons the intensity of sound is attenuated (i.e. it decreases) as it progresses through a medium. Some of this energy is dissipated in the form of heat although there is little appreciable heating of the bulk medium during sonication. The extent of attenuation is inversely related to the frequency. This can be shown by using the example of sound attenuation through pure water. Sound at 118 kHz is reduced to half its original intensity after passing through 1 km of water. At 20 kHz the distance required to achieve the same reduction in intensity is much greater at 30 km (this explains why submarine communications are carried out at low frequencies). These examples indicate that in order to achieve identical intensities in a medium at a given distance from an ultrasonic source using different frequencies, it will be necessary to employ a higher initial power for the source with the higher sound frequency.

1.5 The types of chemical systems affected by power ultrasound

Sonochemistry is mainly concerned with reactions which involve a liquid component within which cavitation can be induced, but since this covers almost all possible chemical situations, sonochemistry has very broad applications. Some typical classes of chemical reaction affected by ultrasonic irradiation are described below.

Homogeneous reactions—involving a single liquid phase

Any system involving a liquid in which bubbles are produced is not strictly homogeneous however in sonochemistry it is normal to consider the original

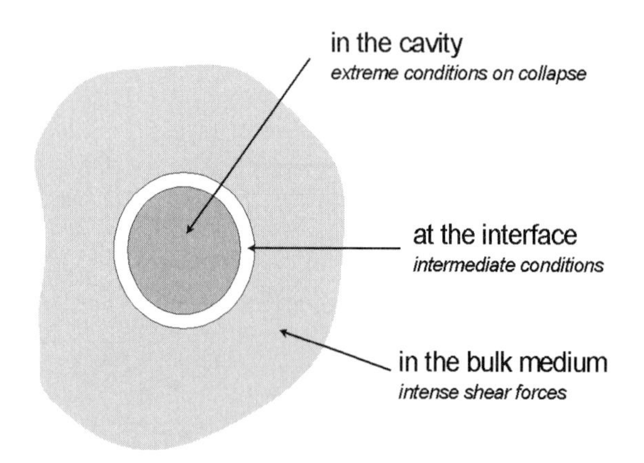

Figure 1.9 Cavitation bubble collapse in a homogeneous medium

state of the system to which the ultrasound is applied. The mechanical and chemical effects of the collapsing bubble will be felt in three distinct regions (Fig. 1.9).

Reactions inside the bubble

The inside of the bubble can be thought of as a high energy microreactor. In order for a chemical to experience the extreme conditions generated inside the cavitation bubble during collapse it must enter the bubble and so should be volatile. The 'concentration' of cavitation bubbles produced by sonication using conventional laboratory equipment is very small and so overall yields in this type of reaction are low. Thus in the sonication of water small quantities of OH˙ and H˙ radicals are generated in the bubble and these undergo a range of subsequent reactions including the generation of H_2O_2 (Scheme 1.3).

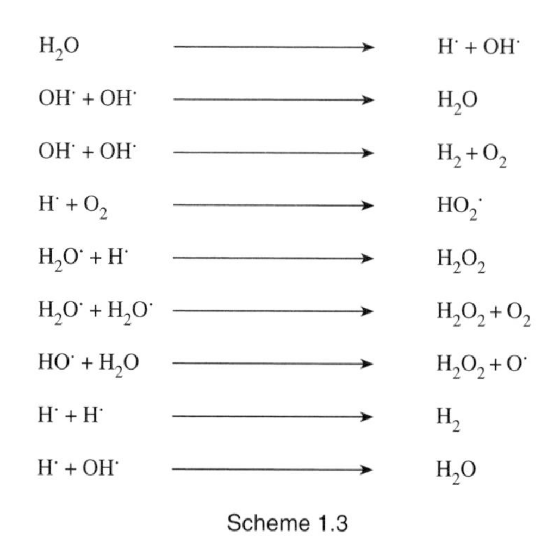

Scheme 1.3

The highly oxidising HO$^{\cdot}$ species can react with other moieties in the bubble or migrate to the bulk solution where they have only transient existence but can be detected chemically [13]. Any species dissolved in the water is clearly going to be subject to chemical reaction with the ultrasonically produced hydrogen peroxide, e.g. if iodide ion is present in solution it will be oxidised and iodine will be liberated (see later). Organic solvents will also slowly decompose on sonication but solvent decomposition normally provides only a minor contribution to any sonochemical reaction taking place in the medium.

A synthetically useful reaction which takes place in the collapsing bubble is the production of amorphous iron from the sonolysis of $Fe(CO)_5$ (0.4 M) in decane under argon [14]. Volatile iron pentacarbonyl enters the bubble and is decomposed during collapse. The fact that an amorphous (rather than crystalline) material is produced confirms that very high temperatures are generated in the bubble and that extreme cooling rates are involved. Conventional production of amorphous iron requires rapid cooling from the vapour to solid state of the order of 10^6 K sec^{-1}. Sonolytic decomposition of iron pentacarbonyl in pentane (a more volatile solvent) yields $Fe_3(CO)_{12}$ rather than the metal, indicating that the cavitation collapse is not so violent in this solvent (Section 1.4). Since this original report, the study of cavitation-induced decomposition of iron and other metal carbonyls has continued and expanded. In the case of molybdenum hexacarbonyl, the product is nanostructured molybdenum carbide which has proved to be an excellent dehydrogenation catalyst [15].

Reactions at or near the bubble/liquid interface

It is tempting to conclude from this that sonochemistry has no effect on materials which cannot enter the bubble. This is not the case as has been shown in a range of reactions during which the very short-lived radicals produced within the bubble migrate to the interface and beyond to undergo reactions with involatile species dissolved in the bulk medium. A good example of this is spin trapping experiments used to identify the radical species produced in the sonolysis of water [12].

Reactions in the liquid immediately surrounding the bubble

The collapse of the bubble also produces very large shear forces in the surrounding liquid capable of breaking the chemical bonding in polymeric materials dissolved in the fluid (Fig. 1.10) [16]. Over the last few years increasing interest has been shown in this procedure since the net result of polymer-chain rupture is a pair of macroradicals, which may recombine randomly (resulting in a reduction in molar mass and possibly leading to a monodispersed system) or act as a radical site on which to polymerise another monomer added to the solution (resulting in block copolymerization).

Small accelerations, in the range 4–15 per cent, have been found for the rate of acid-catalysed hydrolysis of a number of esters of carboxylic acids [17,18]. In the case of methyl ethanoate, the effects (at 23 kHz) were attributed to the

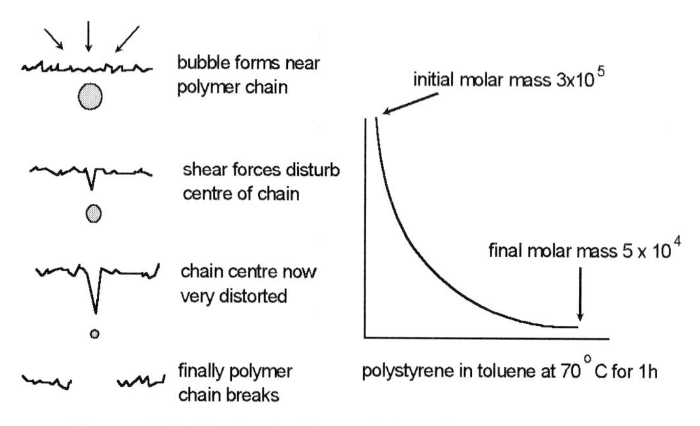

Figure 1.10 Mechanical degradation of dissolved polymers

increased kinetic energy of the molecules induced by the pressure gradients associated with bubble collapse. Similarly, the hydrolysis of the 4-nitrophenyl esters of a number of aliphatic carboxylic acids at 35 °C showed ultrasonically (20 kHz) induced rate enhancements which were all in the range of 14–15 per cent (Scheme 1.4) the activation energy for the hydrolysis of each of the substrates varied considerably with the alkyl substituent (R = Me, Et, *i*-Pr, *t*-Bu) on the carboxylic acid, and so the uniform increase in rate could not be associated with any cavitational heating effect. Here again the modest sonochemical effect was considered to be the result of mechanical effects [19].

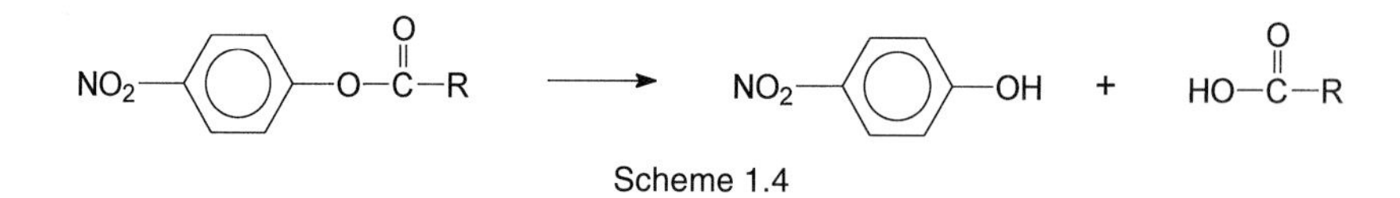

Scheme 1.4

The effect of ultrasonic irradiation on the hydrolysis of 2-chloro-2-methylpropane in mixed aqueous ethanolic solvents of different compositions revealed more evidence for the influence of mechanical effects [20]. The rate enhancement induced by ultrasound (at 20 kHz) was found to increase with an increase in the alcohol content and to decrease as the reaction temperature was raised. A maximum rate increase of 20-fold was observed at 10 °C (in 50 per cent w/w solvent composition). This composition is closely coincident with the structural maximum for the binary ethanol/water solvent system. It is logical to suppose that if the sonochemical enhancement is associated with solvent disruption then the maximum effect would be observed at this composition.

Heterogeneous reactions involving a solid and a liquid

There are two types of reaction involving solid/liquid interfaces: (i) in which the solid is a reagent and is consumed in the process and (ii) in which the solid—often a metal—functions as a catalyst. In any heterogeneous system cavitation which occurs in the bulk liquid phase will be subject to the same conditions as have been described above for homogeneous reactions. There will be a difference, however, when bubbles collapse at or near any interface and this will depend upon the type of solid phase involved.

Cavitation at or near to any large solid surface

In this case the bubble collapse will no longer be symmetrical. The large solid surface hinders liquid movement from that side and so the major liquid flow into the collapsing bubble will be from the other side of the bubble. As a result of this a liquid jet will be formed which is targeted at the surface with speeds in excess of 100 m s^{-1} (Fig. 1.11). The mechanical effect of this is equivalent to high pressure jetting and is the reason why ultrasound is so effective in cleaning. When examined by electron microscopy, surfaces of metals which have been subjected to ultrasonic irradiation reveal 'pitting'. This pitting serves both to expose new surface to the reagents and to increase the effective area available for reaction and is particularly important for the preparation of organometallic intermediates. These are commonly prepared by the direct reaction between a metal and an organic compound. A general problem with such preparations is that the metal surface is easily 'poisoned' by the presence of moisture and other impurities making reaction difficult. In most cases therefore the reagents used for such preparations must be pure and dry and the surfaces of the metal clean and oxide free—as a result these syntheses require special precautions to be taken. Ultrasonic irradiation has made it possible to prepare some organometallic reagents even with technical grade chemicals, conditions unheard of in classical methodology. This is of great economic importance to industry, suggesting that for some syntheses sonication may remove the need to employ super-pure (and thus expensive)

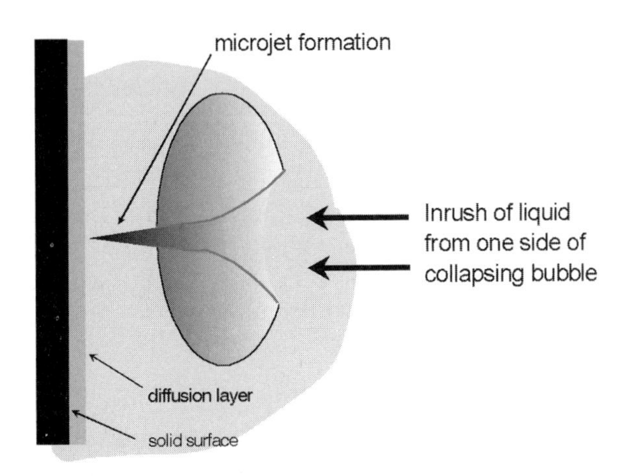

Figure 1.11 Cavitation bubble collapse near a solid surface

Table 1.3 Effect of sonication on a Grignard reaction

Type of ether	Method of agitation	Induction time
Pure and dry	Stirring Sonication	About 6 min Less than 10 s
Reagent grade	Stirring Sonication	Over 3 h About 4 min

chemicals. One example of this is the use of ultrasound for the preparation of Grignard reagents in technical grade diethyl ether [21]. The reaction between magnesium and 2-bromobutane can be ultrasonically induced in under 5 min in this solvent whereas conventional stirring (even with occasional crushing of the metal) takes over 3 h (Table 1.3). Depending upon the conditions used this powerful jet can activate surface catalysis, force the impregnation of catalytic material into porous supports, and generally increase mass and heat transfer to the surface by disruption of interfacial boundary layers.

Cavitation in the presence of a suspended powder

When the solid is particulate in nature cavitation can produce a variety of effects depending on the size and type of material (Fig. 1.12). These include mechanical deaggregation and dispersion of loosely held clusters, the removal of surface coatings by abrasion, and improved mass transfer to the surface. Mechanical deagglomeration is a useful processing aid and is illustrated in the effect of sonication (in a bath) of titanium dioxide pigment in water. A powder sample made up in water consisting initially of agglomerates (volume mean diameter *ca.* 19 μm) was rapidly broken apart (<30 seconds) to provide a limiting size of 1.6-μm particles. Furthermore, the

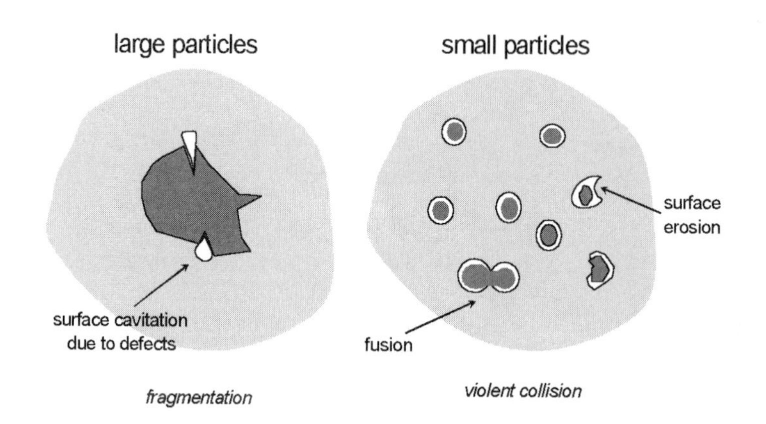

Figure 1.12 Cavitation bubble collapse in a powder suspension

sonicated sample showed no sign of re-agglomeration even after being allowed to stand for a period of 24 hours [22].

A study of the Ullmann coupling reaction has provided evidence that the mechanical effects of surface cleaning coupled with an increase in surface area cannot fully explain the extent of the sonochemically enhanced reactivity. The reaction of 2-iodonitrobenzene to give a dinitrobiphenyl using conventional methodology requires heating for 48 hours and the use of a tenfold excess of copper powder (Scheme 1.5). The use of power ultrasound affords a similar (80 per cent) yield in a much shorter time (1.5 hours) using only a fourfold excess of copper [23]. During these studies it was observed that the average particle size of the copper fell from 87 μm to 25 μm, but this increase in surface area was shown to be insufficient to explain the large enhancement in reactivity (a factor of fifty) produced by ultrasonic irradiation. The studies suggested that sonication assisted in either the breaking down of intermediates and/or the desorption of products from the surface. An additional practical advantage was that sonication prevented the adsorption of copper on the walls of reaction vessels, a common problem when using conventional methodology.

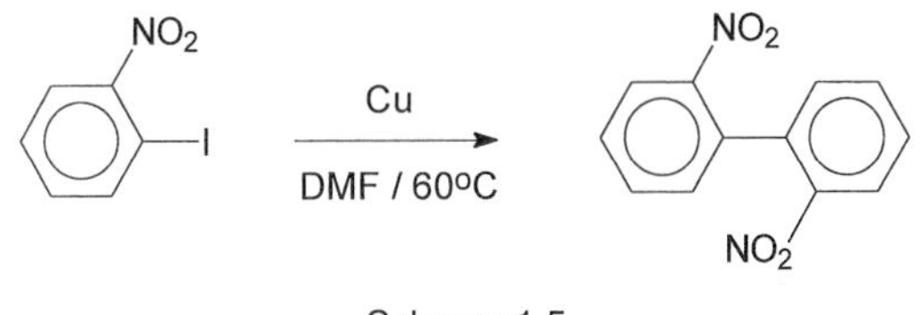

Scheme 1.5

Studies on the effect of sonication on suspended metal powders at 20 kHz in decane at 20 °C have shown that the particles can be forced into such violent collision that fusion can occur [24]. This can be demonstrated for zinc, a soft metal with a low melting point, more significantly for chromium (m.p. 1857 °C) and molybdenum (m.p. 2617 °C), but not for tungsten (m.p. 3410 °C). This led to the conclusion that the collision energy developed was sufficient to raise the temperature of the surfaces in collision to at least 2617 °C but not to 3410 °C—thus giving a measure of the energy involved. In some cases colliding powders can be induced to undergo chemical reaction thus a mixture of powdered copper and sulphur sonicated in hexane generates CuS [25].

Heterogeneous reactions involving immiscible liquids

A problem when dealing with syntheses involving immiscible liquids (e.g. aqueous/organic mixtures) is that the reagents are often dissolved in different phases. Any reaction between these species can only occur in the interfacial region between the liquids and this is a very slow process. Sonication can be used to produce very fine emulsions from immiscible liquids. This is the result of cavitational collapse at or near the interface which causes disruption and impels jets of one liquid into the other to form the emulsion (Fig. 1.13).

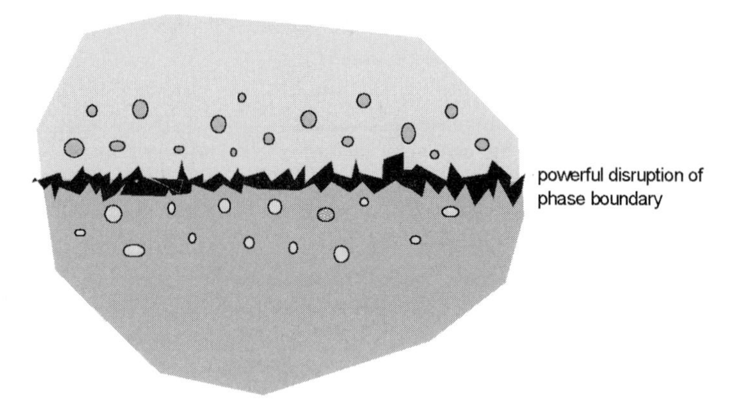

Figure 1.13 Cavitation bubble collapse in a biphasic medium

The normal method of inducing a reaction between species dissolved in different immiscible liquids (usually water and an organic solvent) is through the use of a phase transfer catalyst (PTC) which will bring both reactants into the same, usually organic, phase. There are, however, two drawbacks to the use of such catalysts in that some of the more specialised PTC reagents are expensive, and all the PTCs are potentially dangerous since they can, by their very nature, transfer chemicals from water into human tissue. Sonication of immiscible liquids generates extremely fine emulsions, which result in very large interfacial contact areas between the liquids and a corresponding dramatic increase in the reactivity between species dissolved in the separate liquids. This effect can be used to either replace the need for a PTC or reduce dramatically the quantity required. A good example of this type of application is to be found in the hydrolysis of commercially important oils, fats, and waxes in aqueous NaOH (Scheme 1.6) [26]. The fats are immiscible with the aqueous base and so there is very little contact between fat and NaOH. Traditionally rapid stirring and boiling is used to increase the rate of hydrolysis, but boiling tends to lead to some decomposition and the generation of coloured products. Under sonochemical conditions these hydrolyses can be carried out rapidly and at lower temperatures leading to substantially cleaner products.

$$\begin{matrix} CH_2{-}O{-}COR \\ | \\ CH{-}O{-}COR \\ | \\ CH_2{-}O{-}COR \end{matrix} \xrightarrow{\text{aqueous NaOH}} \begin{matrix} CH_2{-}OH \\ | \\ CH{-}OH \\ | \\ CH_2{-}OH \end{matrix} + 3RCOO\,Na$$

Scheme 1.6

For the chemist the range of applications outlined above encompasses most of the situations encountered in synthesis. In summary we might expect to use ultrasound for a range of applications and perhaps achieve one or more of a number of beneficial effects (Table 1.4).

Table 1.4 Beneficial effects of sonication on chemical reactivity

Accelerate a reaction
Permit the use of less forcing conditions
Make a process more economical by the use of cruder reagents
Revitalise older discarded technologies by enhancing reactivity
Reduce the number of steps required
Initiate stubborn reactions
Reduce any induction period
Enhance catalyst efficiency
Enhance radical reactions

1.6 The uses of ultrasound in conjunction with other techniques

Recent decades have seen the development of a number of new technologies each of which offer the hope of providing the chemist with innovative synthetic routes. Some of them suffer from their own peculiar limitations. For example, photochemistry requires that the chemicals used include a chromophore, and so will not work in highly absorbent solutions. Power ultrasound is not attenuated by a coloured medium and does not require a specific type of functional group to be effective. Sonochemistry certainly falls into the category of a new technology in itself but it has another major advantage for the chemist—it can be used to improve the efficiency of other methodologies. Consider the following examples of the use of ultrasound to help other processes.

High-pressure chemistry (piezochemistry)

Conventional piezochemistry requires specially reinforced, and therefore expensive, reactors. The actual agent volumes involved are often very small (of the order of only a few millilitres for pressures above 1000 bar). In the plant the safety measures are extensive and costly. Ultrasound has been used to reduce the high pressure required for some transformations.

Scheme 1.7 shows the preparation of vanadium hexacarbonyl anion which requires a temperature of 200 °C and 200 atmospheres under conventional conditions. These forcing conditions can be reduced to just 10 °C and only 4.4 atmospheres when performed in the presence of power ultrasound [27]. The use of ultrasound to reduce the forcing conditions required for some transformations is not the only benefit of its use. Sonication does not suffer from the restriction of small sample size.

$$VCl_3 . 3THF \xrightarrow[\text{Na sand}]{\text{CO / THF}} V(CO)_6^-$$

Scheme 1.7

Electrochemistry

These processes often suffer from restrictions in current flow (reduced current efficiency) during the course of the electrolysis because of electrode fouling or hindrance to ionic transport to the electrode surface. Ultrasonic irradiation has been used to great advantage to improve electrochemical processes and the subject has become known as sonoelectrochemistry [28]. The particular advantages include:

(a) degassing at the electrode surface;

(b) disruption of the diffusion layer which reduces depletion of electroactive species;

(c) improved mass transport of ions across the double layer; and

(d) continuous cleaning and activation of the electrode surfaces.

All of these effects combine to provide enhanced yield and improved electrical efficiency.

Catalysis

Solid catalysts suffer from surface deactivation through chemical contamination (poisoning) and passivation during continuous usage. The cleaning effect of ultrasound and its surface activation are important in the enhancement of catalytic reactions. When the catalyst is in the form of a powder or an agglomerate, power ultrasound can also be used to reduce the particle size thereby increasing the available surface area for reaction. This benefit has already been referred to above for an Ullmann reaction (Scheme 1.5) [23].

Supported reagents

When certain chemicals, even enzymes, are adsorbed onto solid supports their reactivity is enhanced over the reagent itself. Ultrasonic irradiation provides a more efficient method of depositing such chemicals onto supports [29]. It also enhances mass transfer to the surface. Consider the case of the substitution of bromide by cyanide (Scheme 1.8). This reaction yields no benzyl cyanide under mechanical agitation in the absence or presence of neutral alumina. However, with alumina present and using ultrasonic irradiation, the substitution product is formed in 64 per cent yield.

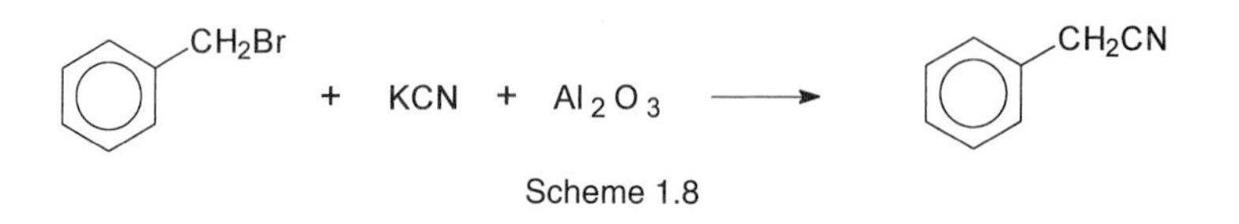

Scheme 1.8

The sonochemically forced impregnation of porous materials has been used in the preparation of a catalyst comprising of 1 per cent (w/w) ruthenium on alumina [30]. The procedure involves depositing $RuCl_3$ from aqueous solution onto alumina (4 mm grain size). The resulting powder is reduced to Ru metal using hydrazine conventionally and in the presence of ultrasound. The resulting Ru metal catalyst prepared under sonochemical conditions was far superior to that prepared conventionally, in that there was a greater penetration of the metal into the support with no metal close to the surface.

1.7 The generation of ultrasound

A transducer is the name for a device capable of converting one form of energy into another, a simple example being a loudspeaker which converts electrical energy to sound energy. Ultrasonic transducers are designed to convert either mechanical or electrical energy into high frequency sound and there are three main types: gas driven, liquid driven, and electromechanical.

Gas-driven transducers

These are, quite simply, whistles with high frequency output (the dog whistle is a familiar example). The history of the generation of ultrasound via whistles dates back 100 years to the work of F. Galton who was interested in establishing

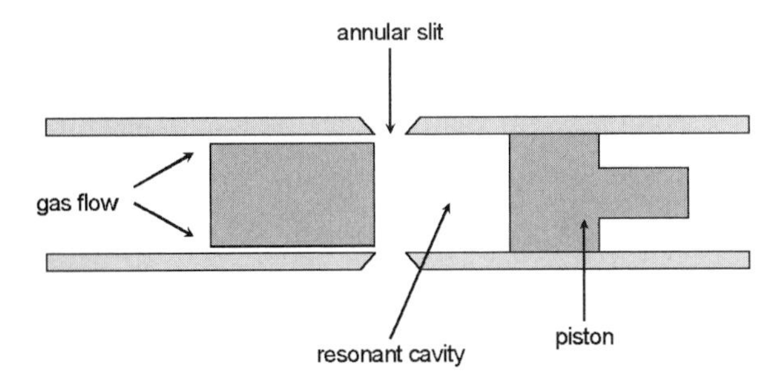

Figure 1.14 Schematic diagram of Galton's whistle

the threshold levels of human hearing [31]. He produced a whistle which generated sound of known frequencies, and was able to determine that the normal limit of human hearing is around 18 kHz. Galton's whistle was constructed from a brass tube with an internal diameter of about two millimetres (Fig. 1.14), and was operated by passing a jet of gas through an orifice into a resonating cavity. On moving the plunger the size of the cavity could be changed to alter the 'pitch' or frequency of the sound emitted. An adaptation of this early principle is to be found in some dog whistles which have adjustable pitch.

An alternative form of gas-generated ultrasound is the siren. When a solid object is passed rapidly back-and-forth across a jet of high pressure gas it interferes with the gas flow and produces sound of the same frequency at which the flow is disturbed. A siren can be designed by arranging that the nozzle of a gas jet impinges on the inner surface of a cylinder through which there are a series of regularly spaced perforations. When the cylinder is rotated the jet of gas emerging from the nozzle will rapidly alternate between facing a hole or the solid surface. The pitch of the sound generated by this device will depend upon the speed of rotation of the cylinder.

Neither type of transducer has any significant chemical application since the efficient transfer of acoustic energy from a gas to a liquid is not possible (see above). However, whistles are used for the atomization of liquids.

The conventional method of producing an atomized spray from a liquid is to force it at high velocity through a small aperture. (A typical domestic example being a spray mist bottle for perfume.) The disadvantage in the design of conventional equipment is that the requirement for a high liquid velocity and a small orifice restricts its usage to low viscosity liquids, and these atomizers are often subject to blockage at the orifice.

Figure 1.15 shows a schematic gas driven atomizer. The system comprises of an air or gas jet which is forced into an orifice where it expands and produces a shock wave. The result is an intense field of sonic energy focused between the nozzle body and the resonator gap. When liquid is introduced into this region it is vigorously sheared into droplets by the acoustic field. Air by-passing the resonator carries the atomized droplets downstream in a fine soft plume shaped spray. The droplets produced are small and have a low

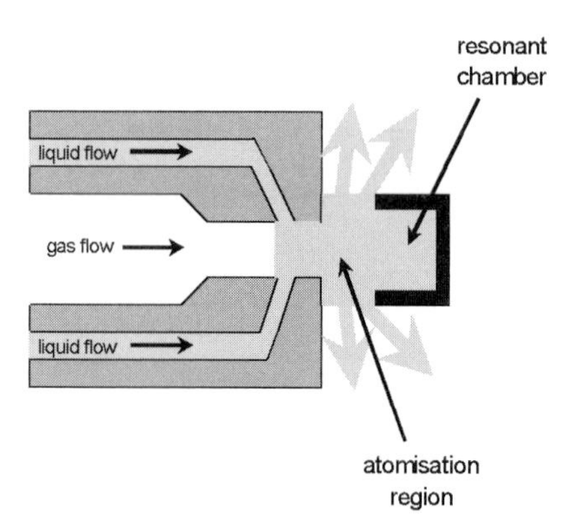

Figure 1.15 Schematic diagram of a compressed air spray atomiser

forward velocity. Atomized water sprays have many uses including dust suppression in industry and humidifiers for horticultural use under glass.

Liquid-driven transducers

In essence this type of transducer is a 'liquid whistle' and generates cavitation via the motion of a liquid rather than a gas. Process material is forced at high velocity by the homogeniser pump through a special orifice from which it emerges as a jet which impacts upon a steel blade (Fig. 1.16). There are two ways in which cavitational mixing can occur at this point.

Firstly, through the Venturi effect as the liquid rapidly expands into a larger volume on exiting the orifice, and, secondly, via the blade which is caused to vibrate by the process material flowing over it. The relationship between orifice and blade is critically controlled to optimise blade activity. The required operating pressure and throughput is determined by the use of different sizes and shapes of the orifices, and the velocity can be changed to achieve the necessary particle size or degree of dispersion. With no moving

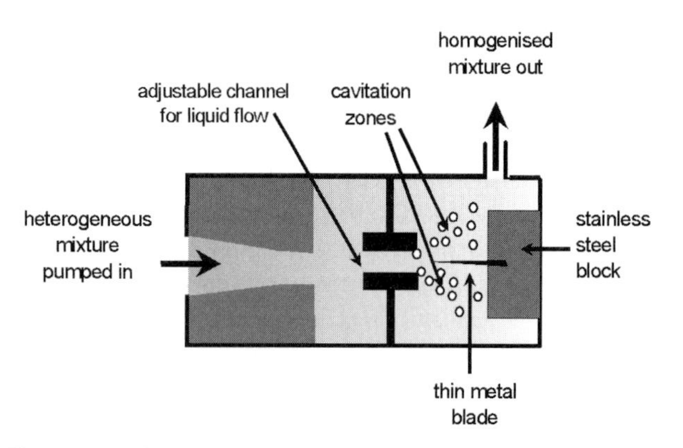

Figure 1.16 Schematic diagram of a liquid whistle homogeniser

parts, other than a pump, the system is rugged and durable. When a mixture of immiscible liquids is forced through the orifice and across the blade, cavitational mixing produces extremely efficient homogenisation.

Equipment employing this principle has been in existence for nearly forty years. The main user has been the food industry where the homogenisation of immiscible liquids has always been important; however, more recently other industries have turned to this simple device for processing (Table 1.5).

Table 1.5 Industrial applications of ultrasonic mixing

Food	Textiles	Others
Soups and sauces	Dispersed dyestuffs	Blending polymers
Ketchup	Thread lubricants	Cosmetic formulation
Salad creams	Printing thickeners	Precipitation
Mayonnaise	Starch sizing solutions	
Fruit juices		

Electromechanical transducers

The two main types of electromechanical transducers are based on either the piezoelectric or the magnetostrictive effect. The most commonly used of which are piezoelectric transducers, generally employed to power the bath and probe type sonicator systems. Although more expensive than mechanical transducers, electromechanical transducers are by far the most versatile.

Magnetostrictive transducers

Historically magnetostrictive transducers were the first to be used on an industrial scale to generate high-power ultrasound. These are devices that use an effect found in some materials, e.g. nickel, which reduce in size when placed in a magnetic field and return to normal dimensions when the field is removed (magnetostriction). When the magnetic field is applied as a series of short pulses to a magnetostrictive material it vibrates at the same frequency. In simple terms, such a transducer can be thought of as a solenoid in which the magnetostrictive material (normally a laminated metal or alloy) forms the core with copper wire winding. To avoid magnetic losses two such solenoids are wound and connected in a loop (Fig. 1.17).

The major advantages of magnetostrictive systems are that they are of an extremely robust and durable construction and provide very large driving forces. This makes them an attractive proposition for heavy-duty industrial processing. There are, however, two disadvantages, firstly the upper limit to the frequency range is 100 kHz, beyond which the metal cannot respond fast enough to the magnetostrictive effect, and secondly the electrical efficiency is less than 60 per cent with significant losses emerging as heat. As a result of the second of these problems all magnetostrictive transducers subject to

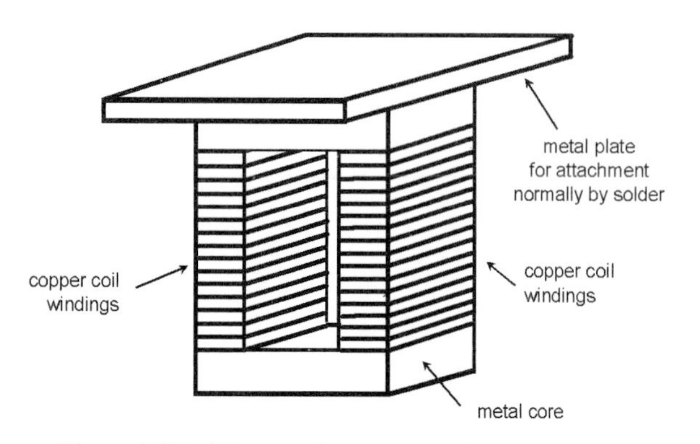

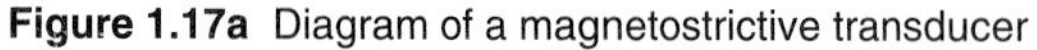

Figure 1.17a Diagram of a magnetostrictive transducer

Figure 1.17b Photograph of a magnetostrictive transducer

extended use are liquid cooled. This has meant that piezoelectric transducers (see below) which are more efficient and operate over a wider frequency range are generally considered to be the better choice in sonochemistry, especially in laboratory situations. However, now that a range of industrial applications for sonochemistry are under consideration, particularly those requiring heavy-duty continuous usage at high operating temperatures, the magnetostrictive transducer is coming back into consideration.

Many improvements in the operating efficiency of this type of transducer have been made all of which are based on finding a more efficient magnetostrictive core. The original nickel-based alloys have been replaced by more electrically efficient cobalt/iron combinations and, more recently, aluminium/iron with a small amount of chromium. One of the latest developments in magnetostrictive technology has been the introduction of a new material called TERFINOL-D. This is an alloy of the rare earths terbium and dysprosium with iron, which is zone refined to produce a material almost in the form of a single crystal. It can be produced in various forms, rods, laminates, tubes etc., and has several major advantages over the more conventional alloys used. A magnetostrictive transducer based on this material can generate more power than a conventional piezoelectric transducer, it is compact (about 50 per cent smaller) and lighter than other magnetostrictives. It does have the same problem as other such devices in that it has an upper limit of frequency response—in this case 70 kHz.

Piezoelectric transducers

The most common types of transducer used for both the generation and detection of ultrasound employ materials which exhibit the piezoelectric effect, discovered over a century ago [32]. Such materials have the following two complementary properties:

1. The direct effect—when pressure is applied across the large surfaces of the section a charge is generated on each face equal in size but of opposite sign. This polarity is reversed if tension is applied across the surfaces.

2. The inverse effect—if a charge is applied to one face of the section and an equal but opposite charge to the other face, then the whole section of crystal will either expand or contract depending on the polarity of the applied charges. Thus, on applying rapidly reversing charges to a piezoelectric material fluctuations in dimensions will be produced. This effect can be harnessed to transmit ultrasonic vibrations from the crystal section through whatever medium with which it is in contact.

Quartz was the piezoelectric material originally used in devices such as the very early types of ASDIC underwater ranging equipment. Quartz is not a particularly good material for this purpose because of its mechanical properties, it is somewhat fragile and difficult to machine. Modern transducers are based on ceramics containing piezoelectric materials. These materials cannot be obtained as large single crystals and so, instead, they are ground with binders and sintered under pressure at above 1000 °C to form a ceramic. The crystallites of the ceramic are then aligned by cooling from above their ferroelectric transition temperature in a magnetic field. Such transducers can be produced in different shapes and sizes. Nowadays the most frequently employed piezoceramic contains lead zirconate titanate (commonly referred to as *pzt* where the *p* represents plumbum—the chemical term for the element lead—and the *z* and *t* are initials from the name of the salts). The most common form is a disc with a central hole. In a power transducer, it is normal practise to clamp two of these piezoelectric discs between metal blocks which serve both to protect the delicate crystalline material and to prevent it from overheating by acting as a heat sink. The resulting 'sandwich' provides a durable unit with doubled mechanical effect (Fig. 1.18). The unit is generally one half wavelength long (although multiples of this can be used). The peak

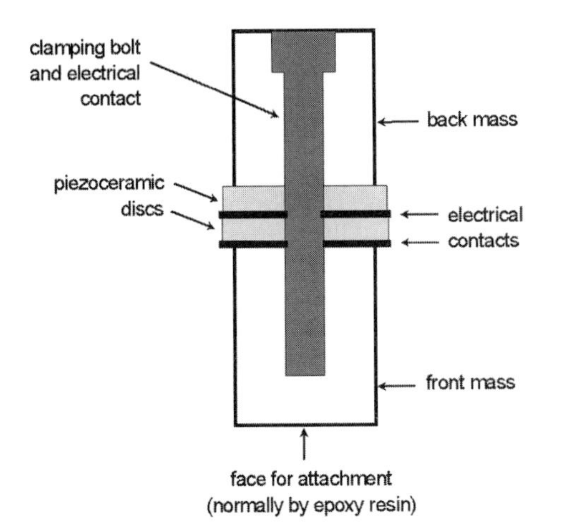

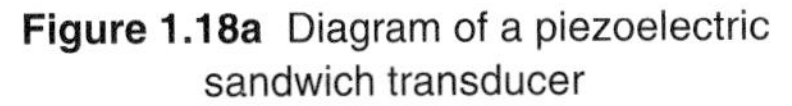
Figure 1.18a Diagram of a piezoelectric sandwich transducer

commercial 40kHz sandwich transducer
ceramic disk 3.7cm diameter, hole 1.5cm diameter, thickness 0.6cm

Figure 1.18b Photograph of a piezoelectric sandwich transducer

to peak amplitudes generated by such systems are normally of the order of 10–20 microns and they are electrically efficient. Generally piezoelectric devices must be cooled if they are to be used for long periods at high temperatures because the ceramic material will degrade under these conditions.

Such transducers are highly efficient (>95 per cent) and, depending on dimensions, can be used over the whole range of ultrasonic frequencies from 20 kHz to many MHz. They are the exclusive choice in medical scanning which uses frequencies above 5 MHz.

1.8 The cavitation threshold

When very low power ultrasound is passed through a liquid and the power is gradually increased, a point is reached at which the intensity of sonication is sufficient to cause cavitation in the fluid. It is only at powers above the cavitation threshold that sonochemistry can occur because only then can the great energies associated with cavitational collapse be released into the fluid. In the medical profession, where the use of ultrasonic scanning techniques is widespread, keeping scanning intensities below the cavitation threshold is of vital importance. As soon as the irradiation power used in the medical scan rises above this critical value cavitation is induced and, as a consequence, unwanted even possibly hazardous chemical reactions may occur in the body. Thus, for both chemical and medical reasons there is a considerable drive towards the determination of the exact point at which cavitation occurs in liquid media, particularly in aqueous systems. There are essentially three methods which can be used for the determination of the cavitation threshold relying on either the initial detection of bubbles in the medium, the emission of light (sonoluminescence), or the identification of a chemical reaction which occurs only in the presence of cavitation.

Determination by the detection of bubbles

Bubbles are extremely efficient absorbers of ultrasound and this provides the basis for a sensitive measurement of the threshold. A low intensity ultrasonic source is arranged with a receiver in a liquid such that the absorption (attenuation) of sound in that liquid can be assessed. Power ultrasound from a separate source can now be applied to the liquid and the attenuation continuously monitored as the sonication power is increased. The onset of cavitation is registered when the attenuation suddenly leaps to a higher value due to the presence of cavitation bubbles in the fluid.

Determination through the detection of sonoluminescence

Sonoluminescence is the emission of a weak light from cavitating fluids. This emission can be engendered by power ultrasound operating above the cavitation threshold. Sonoluminescence is outside of the scope of this book but a comprehensive review has been published recently. The detection of the onset of sonoluminescence for the determination of the cavitation threshold has been used successfully by Crum and co-workers [33], who have described in detail the experimental apparatus required. This includes both a 'fluid management system' to provide ideal conditions in the liquid to be sonicated and a schematic diagram showing the circuitry used for the generation of cavitation.

Determination through the initiation of chemical reactions

For the practising sonochemist the above systems are perhaps too 'physics' oriented, when the main interest for them is the onset of cavitation-induced chemistry. The point at which cavitation induces chemical reaction can be used as a true measure of the cavitation threshold which occurs in an actual reaction vessel, i.e. one which is later to be used for sonochemical experiments. Here we will look at two different types of reaction which have been used as chemical probes for the onset of cavitation. By definition these must be particularly sensitive in detecting low levels of radical species. Other, more routine, dosimeters also exist and these are used for the measurement of the amount of cavitation induced by prolonged ultrasonic irradiation. These will be discussed later in this chapter.

Spin trapping and esr identification of radicals

When cavitation is induced in a liquid, one possible result is the production of free radicals. The hydroxy and hydrogen radicals produced by cavitation in aqueous media are short-lived species which can often be identified by spin trapping and subsequent esr (electron spin resonance) examination of the spin adduct formed [13]. Spin traps are usually nitroso or nitrone derivatives and the adducts are the longer-lived nitroxide radicals [31]. For the detection of hydroxyl radicals, 5,5-dimethylpyrroline-*N*-oxide (DMPO) is particularly useful, while for hydrogen atoms, α-(4-pyridyl-1-oxide)-*N*-tertbutylnitrone (4-POBN) is preferable because of the longer lifetime of the spin adduct (Scheme 1.9). These methods have been used to study aqueous solutions of amino acids, peptides, DNA bases, sugars, nucleosides, nucleotides, surfactants, and mixtures of water and alcohols [32].

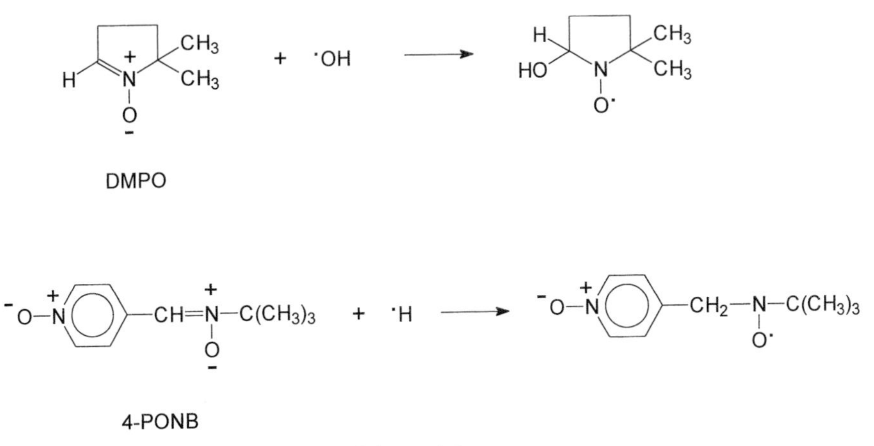

Scheme 1.9

Using a fluorescence technique

Terephthalic acid (TA) has often been used as a chemical dosimeter for ionising radiation and functions effectively as a trap for hydroxy radicals [33]. In aqueous solution TA produces terephthalate anions which react with hydroxyl radicals $HO^{\cdot}$ to produce highly fluorescent hydroxyterephthalate ions (HTA) (Scheme 1.10) [34]. This reaction provides a very sensitive method of measuring the hydroxyl radicals produced by the sonolysis of water and is particularly useful in determining the cavitation threshold [35].

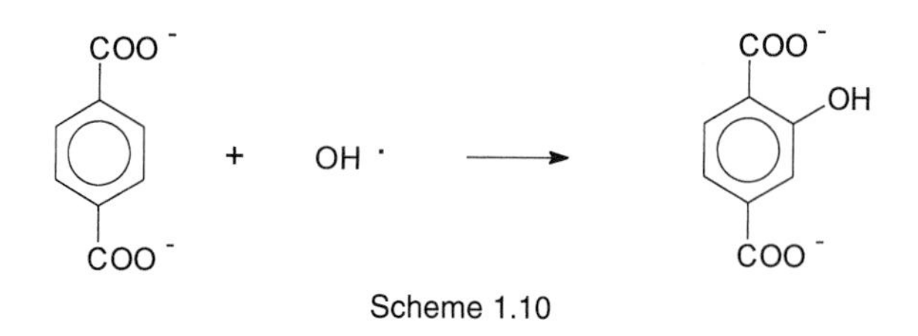

Scheme 1.10

The product of hydroxy radical reaction with TA is hydroxyterephthalic acid (HTA) which can be estimated by fluorescence measurements at an analysing wavelength of 425 nm with an excitation wavelength of 315 nm. Very good linearity can be achieved for fluorescence against concentration for HTA down to 3×10^{-8} M, whereas the lower limit for TA is only 2×10^{-3} M. Thus with this system it is possible to detect a 10^{-5} M fraction of TA transformed into HTA by $HO^{\cdot}$ radicals.

An ultrasonic exposure system for the determination of cavitation threshold

In diagnostic ultrasound the onset of cavitation must be avoided, and so great efforts are made to set up a well-controlled exposure environment for cavitation threshold measurements in a fluid. With the aid of such an environment the onset of cavitation can be recorded using one of the sensitive dosimeters already described. The exposure system described below has two advantages. First, it generates a well-characterised acoustic beam that is reproducible. Second, it delivers both constant wave (CW) and pulsed ultrasound, the latter being the type used in medical scanning. The range of intensities generated by this equipment simulates most therapeutic and diagnostic applications of medical ultrasound. The onset of cavitation is signalled by the production of free radicals and two methods for their detection are described: (a) by esr spin trapping techniques and (b) by fluorescence monitoring [13,39].

The exposure tank is fabricated from 2.5-cm thick sheets of Plexiglas and measures 18 × 30 × 22 cm (Fig. 1.19). At one end an acoustic transducer is mounted, with an absorber positioned directly opposite across the tank to minimise reflections which might result in standing waves. The absorber chosen was castor oil since its acoustic impedance matches that of water and it has a high attenuation coefficient. An acoustic window, made from a thin

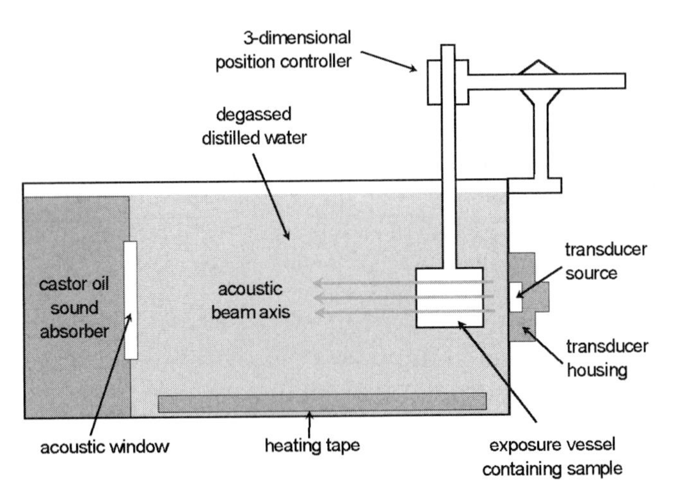

Figure 1.19 Ultrasonic exposure tank

household plastic film (e.g. cling film), separates the oil from the water. Better acoustic absorbing materials are now available, e.g. a silicone rubber compound filled with microballoons [40]. A constant temperature was maintained in the exposure tank to avoid thermal gradients which would disturb the acoustic beam. Bubbles were also eliminated by using degassed water (boiling distilled water under a vacuum then cooling it in a sealed vacuum bottle makes sufficiently high quality degassed water).

The reaction cell resembles an inverted T, with a glass cylinder 3.3 cm in diameter and 2.5 cm in length forming the base. The stem is used for mounting and filling the cell which has a sample volume of 20 cm. Both ends of the tube are sealed with an acoustically transparent film (again cling film). Gas was bubbled through the sample prior to sonication but this was stopped before the exposure began. The leading edge of the tube was 6 cm from the surface of the transducer. Before each exposure the tube was centred and aligned with the axis of the acoustic beam.

A note of warning must be sounded at this point. Controlling the exposure field does not guarantee that the sonochemical results themselves will be reproducible. Using the above apparatus the acoustic field was stable and reproducible but the sonochemical results were not. Cavitation at low-level power, i.e. just above the threshold, is an elusive phenomenon since, apart from factors which affect the exposure field, it also depends on the state of the liquid itself [41].

1.9 Ultrasonic power measurement

The two commonly used sources of ultrasound in the chemical laboratory are the ultrasonic cleaning bath and the probe system. The bath system transmits energy into water contained in the bath, so that acoustic energy can be transferred to any reaction flask dipped into the water. The bath is generally low in energy and of fixed power. The more powerful probe system, which is of variable power, transmits acoustic energy directly into a reaction mixture via a wave guide (usually made of titanium) directly attached to the power transducer. These systems will be dealt with in detail in subsequent chapters.

Any paper reporting sonochemical results would be expected to specify the make and model of the irradiation source employed, and this is normally so. Unfortunately, the power input to the reacting system is sometimes reported only as the quoted maximum rating for the equipment. A 500-W unit does not mean that the instrument delivers 500 W of ultrasonic power into any system when operating at its maximum setting. In order to record the actual power input to a reaction some standard method of measurement is required. These methods may be divided into physical and chemical dosimetry measurements [42].

Physical measurements

The measurement of vibrational amplitude for a probe system

This is the direct measurement of the amplitude at the vibrating tip of the horn, i.e. the actual mechanical motion being transmitted to the chemical reaction in which the probe is immersed. This will give a parameter that should be directly proportional to the acoustic power. The measurement of amplitude can be achieved simply by observing the ultrasonic vibration using a metallurgical microscope with a calibrated eyepiece. Since most transducers will generate amplitudes of at least 10 microns, quite accurate measurements can be made. A small spot of aluminium paint is placed on the surface of the probe and a single metallic fleck is focused in the graticule. On turning on the power the rapidly vibrating fleck appears as a 'smear', the length of which is the amplitude of vibration. This method of measurement cannot be easily made when the probe is in use during a sonochemical experiment. An electromechanical method is available, however, which provides continuous monitoring with a display while the probe is in use. The alternating stress in a resonating horn is at a maximum in its centre (this is a stationary point, cf. the middle of a stretched elastic band). If a strain gauge is bonded across this point then the output from this will be proportional to the displacement or amplitude of vibration. This output signal can be displayed on a meter. The meter can then be calibrated by the use of a microscope as described above. Vibrational amplitude is not an absolute measure but it does offer a very sensitive monitor for acoustic changes during a sonochemical reaction. Its major drawback is that a strain gauge is a somewhat fragile instrument; however, amplitude gives an indication of the acoustic power output rather than the electrical power into the transducer. Using this method it is only necessary to calibrate the meter once, since any subsequent change in transducer amplitude due to loading will be accompanied by a proportional change in strain in the system. The meter will thus follow any induced changes of amplitude which occur either as a result of power input or changes in properties of the sonicated reaction.

Measurement of the real electrical power to the transducer

A wattmeter can be used to measure electrical power to the transducer. When the system is operating in a chemical reaction it is possible to derive the transmitted acoustic power by measuring electrical power to the transducer when the system is running in air (i.e. the probe tip is not immersed in a reaction mixture—a situation which is technically referred to as a system operating without a load) and then when it is loaded (i.e. dipped into the system). The transmitted electrical power is the difference between these values. From this the acoustic power entering the reaction can be calculated if the acoustic transfer efficiency is known [43].

Calorimetric measurement of ultrasonic power entering a reaction

Of the methods available to measure the amount of ultrasonic power entering a sonochemical reaction the most common is calorimetry, which involves a measurement of the initial rate of heating produced when a system is irradiated by power ultrasound. In a pure liquid, one might assume that almost all the mechanical energy produces heat and so, via calorimetry, some estimate of the acoustic power entering the system can be obtained. This is a simple method of estimating the overall acoustic power entering a reaction input even when a thermally insulated vessel is not employed. For the system under study the temperature (T) is recorded against time (t), at intervals of a few seconds, using a thermocouple placed in the reaction itself. From the T versus t data, the temperature rise at zero time, ($\mathrm{d}T/\mathrm{d}t$), can be estimated by either constructing a tangent to the curve at time zero, or by curve-fitting the data to a polynomial in t. The ultrasonic power actually entering the system can then be obtained by substituting the value of ($\mathrm{d}T/\mathrm{d}t$) at time zero (obtained from either method) into eqn (1.1).

$$\text{Power} = (\mathrm{d}T/\mathrm{d}t)c_{\mathrm{p}}M \tag{1.1}$$

where c_{p} = heat capacity of the solvent (J $\mathrm{kg^{-1}\ K^{-1}}$) and M = mass of solvent used (kg). If this Power (in Watts) is dissipated into the system from a probe tip with an Area measured in $\mathrm{cm^2}$, then the intensity of power (in W $\mathrm{cm^{-2}}$) produced by the source of ultrasound is given by eqn (1.2).

$$\text{Intensity} = \text{Power/Area} \tag{1.2}$$

Chemical measurements

A chemical dosimeter monitors only the production of a chemical species through sonochemistry. In a sense this is the true dosimeter required by a chemist, but it does not reflect the total energy entering the system since not all events occurring inside the collapsing bubble give rise to observable chemical products in the bulk liquid. In the case of the sonolysis of water some of the radicals (e.g. $H^{\cdot}$ and $OH^{\cdot}$) recombine to regenerate water.

A few chemical dosimeters are in common usage.

The iodine dosimeter (oxidation of I^- to I_2)

The iodine dosimetry method is based upon the fact that sonication of water produces H_2O_2 and this reacts quickly with I^- in solution to liberate molecular iodine. Thus, when aliquots of standardised H_2O_2 (of the order of 1×10^{-5} M dm^{-3}) are added to KI solution, fixed quantities of iodine will be liberated, and the yield of iodine can be estimated by the increase in absorbance of this species at 355 nm on a UV/visible spectrophotometer. With the aid of such a Beer Lambert plot it is possible to calculate the concentration of H_2O_2 which gives rise to a particular absorbance of iodine and hence the yield of peroxide generated from a particular sonolysis of water. The only difficulty with this method is that the quantity of peroxide liberated on insonation of aqueous KI is quite low [44] and so great care should be taken over the cleanliness of equipment. The calibration plot can then be used to estimate the amount of peroxide generated by various sonication systems simply by putting 5 per cent aqueous KI in the reaction vessel and measuring the development of iodine absorbance with sonication time. In this way comparisons can be made between different pieces of sonochemical apparatus. As we will see later, for consistent results it is important to carry out sonications with the various components of the reactor in the same configuration. In addition it is helpful if the same flask is used throughout these measurements—particularly for reactions involving ultrasonic baths—since this will help to ensure that the same amount of energy is transferred into the reaction through the glass base of the flask.

The rate of iodine liberation from KI solution can be enhanced by a factor of five in the presence of a small amount of CCl_4 (the so-called Weissler solution), and this method can be used to demonstrate the process (see Chapter 2). The enhancement is the result of sonochemical generation of additional radicals from the haloalkane, which liberate iodine more readily from the KI. The results from this method are thus a reflection of the total number of radicals produced during insonation and are not restricted to those generated from water alone [45].

The Fricke dosimeter (oxidation of Fe^{2+} to Fe^{3+})

Aqueous sonochemistry can be thought of as similar in some ways to radiation chemistry. One such similarity is in the oxidation of an air-saturated solution of Fe^{2+} to Fe^{3+}, a system analogous to the classical Fricke dosimeter in radiation chemistry. In 1950 Miller [46] studied this oxidation induced by ultrasound of frequency 500 kHz. The yield was found to be dependent upon the length of irradiation up to 10 minutes duration, beyond which non-linearity of yield was encountered. If, however, at the end of the linear period the solution was resaturated with air and then further irradiated, a linear concentration/time was re-established with identical slope to that obtained previously. This observation suggests that the non-linearity is due to ultrasonically induced degassing of the solution. Clearly estimates of energy input must therefore be made based upon initial concentration changes of the Fe^{3+}

ions generated. The oxidation occurs as the result of an indirect process. Cavitational collapse first generates hydroxy radicals and these react with Fe^{2+} to yield hydroxide ion and Fe^{3+} (eqn 1.3).

$$Fe^{2+} + {}^{\cdot}OH \rightarrow Fe^{3+} + {}^{-}OH \qquad (1.3)$$

A typical solution would be composed of ferrous ammonium sulphate (1 mM) and sodium chloride (1 mM) dissolved in sulphuric acid (0.5 M), and the radiation dose can be calculated based on the increase in absorbance of the Fe^{3+} ions at 304 nm [47] (extinction coefficient 2194 M^{-1} cm^{-1}).

Terephthalate dosimeter

The sensitivity and method of use of this system have already been referred to in terms of the detection of cavitation threshold (Section 1.8 and Scheme 1.10). McLean and Mortimer [48] have studied the variations in $HO^{\cdot}$ free radical production during the sonication of aqueous solutions at different powers at 970 kHz. Typically the results showed a threshold power for radical production, after which there is a linear correlation with acoustic power indicating its potential for use as a dosimeter. The terephthalate chemical dosimeter can be prepared by dissolving terephthalic acid (TA) 1.5×10^{-3} mol $litre^{-1}$ and NaOH 5×10^{-3} mol $litre^{-1}$ in a phosphate buffer at pH = 7.4. The fluorescence was measured as in Section 1.8. Mason *et al.* reported the response of the TA dosimeter with different ultrasonic sources and frequencies. They employed probe systems operating at 20, 40, or 60 kHz [49]. Ultrasonic power measurements were monitored using the calorimetric method. It was reported that under constant sonication conditions the measured fluorescence was directly proportional to exposure time. Within the power ranges studied, the yield of $HO{\cdot}$ radicals was proportional to the power input, and at constant power, the fluorescence yield increased from 20 to 60 kHz, i.e. it appears that radicals are produced more efficiently at higher frequencies.

Nitrophenol dosimeter

Free radicals can also react with aromatic systems and this type of reaction has been used as a method of dosimetry in sonochemistry. Hydroxy radicals generated sonochemically in an aqueous solution containing 4-nitrophenol react to produce nitrocatechol (Scheme 1.11).

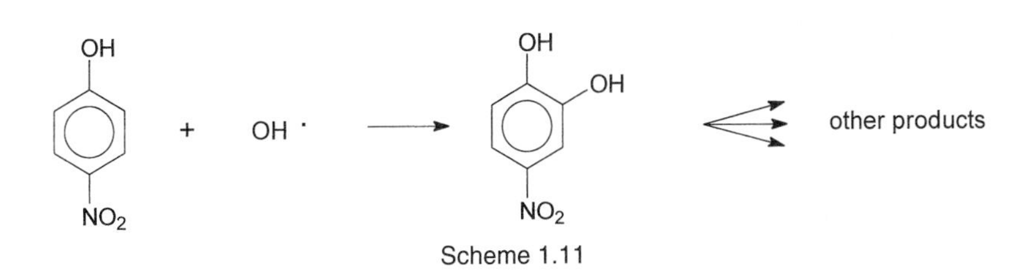

Scheme 1.11

Aqueous 4-nitrophenol has an ultraviolet maximum at 401 nm and the hydroxylated product at 512 nm. It is therefore possible to use this reaction as a dosimeter for hydroxy ions in a similar way to that used in the terephthalic acid system [50]. In this case a calibration curve is required for the product 4-nitrocatechol (from zero to about 2×10^{-5} M dm^{-3}). The nitrophenol dosimeter is an aqueous solution (10^{-4} M dm^{-3}) at pH 5, adjusted using phosphoric acid. The main drawback of this method is that the reactive hydroxy radical can also attach to the first formed product nitrocatechol, and hence lead to other materials with different spectra. For this reason the method tends to be less accurate for high-powered ultrasound over extended monitoring periods.

In conclusion it is important to note that the above chemical dosimeters do not measure the same effects. The TA and nitrophenol dosimeters are specific for $HO^{\cdot}$ radicals, while the others are more general, thus both I^- and Fe^{2+} can also be oxidised by $HO_2^{\cdot}$, H_2O_2, or indeed any other species, and such processes do not occur at the same rate (e.g. the rate of production of I_2 from I^- oxidation and the formation of H_2O_2 in water can be monitored independently and are not the same). Chemical dosimeters are strongly frequency dependent, thus the production of iodine in air-saturated KI solutions is six times faster at 514 kHz than at 20 kHz [51]. They are also strongly dependent on experimental conditions especially with respect to the gas content.

1.10 Types of sonochemistry equipment

In subsequent chapters the various types of equipment which can be used for sonochemistry will be explored in depth. In this brief section the types of instrumentation currently available for sonochemistry will be identified. The versatile ultrasonic bath (Chapter 2) and the more powerful probe system (Chapter 3) are the most commonly available to the chemist. The source of power ultrasound for both is generally the piezoelectric transducer and each suffers from the same disadvantage that they operate at a fixed frequency depending on the particular transducer employed. For most commercial probe systems this frequency is 20 kHz and for baths it is 40 kHz. To study sonochemistry at different frequencies it is best to purchase separate systems tailored to individual requirements. It is also possible to construct a package which includes a multiple frequency amplifier from which a set of different horns can be driven.

These two systems are routinely used in the batch mode. Some laboratory systems offer the possibility of processing quite large volumes by sonication of a part of the reacting mixture as it flows through a sonication cell (flow systems are studied in Chapter 4). This offers the potential for the processing of larger volumes and is probably the most likely way that sonochemistry could be applied in industry.

1.11 The safety of sonochemical equipment

There is perhaps a temptation to think that if you cannot hear a noise it cannot be harmful. This is certainly true of sound in the audible range because in that situation zero ear response means very low intensity indeed. With ultrasound, however, much more care must be taken. It is important to note that most ultrasonic devices when irradiating a liquid medium do give rise to some audible sound. This is because the sound spectrum emitted from a sonicated reaction is not composed entirely of the single transducer resonance but includes some sub-harmonics which give rise to very audible sound levels. Ultrasound is attenuated rapidly in air but sources of ultrasound such as probes start at such a high intensity that there is a distinct possibility that long-term exposure without ear protection might result in damage. Consideration must therefore be given to the health and safety aspects associated with the use of ultrasound. Permitted exposure limits do vary from country to country but generally they lie between 85 and 90 dBA over an 8-hour period. A dBA is a technical exposure term representing an average effect of sound pressure as perceived by the ear, i.e. not the actual distribution of sound pressures from the sound spectrum. Sound exposure meters are generally calibrated in these units. Like radioactivity shorter exposure times permit higher dBA levels. The simplest course of action is that those laboratories which employ sonicators should refer to the latest current industrial practice in their own country. There are three ways in which workers can be protected against radiated noise:

1. Generally, any person working with ultrasonic equipment should wear either acoustic ear muffs or in the ear protectors. These are generally available for use in noisy industrial environments. The simple ultrasonic cleaning bath is not normally sufficiently noisy to warrant more than this simple precaution.

2. It is good practise to isolate ultrasonic equipment in laboratories where access is limited—a closed door is a reasonable sound barrier.

3. It is recommended that an acoustic screen is placed around the ultrasonic equipment—particularly in the case of powerful probe systems. This can take the form of a box lined with a sound absorbing material, say expanded polystyrene or polythene bubble wrapping. In the chemical laboratory a fume cupboard can be partially sound proofed in this way. A neat solution for the teaching laboratory is a sound-proof box mounted on wheels so that it can be moved around to wherever it is needed.

References

1. M. J. W. Povey, and T. J. Mason, *Ultrasound in food processing* (1998) Thomson Science, London.
2. B. Brown, and J. E. Goodman, *High intensity ultrasonics* (1965) ILiffe Books, London.
3. J. R. Frederick, *Ultrasonic engineering* (1965) John Wiley, London.
4. O. V. Abramov, *High-intensity ultrasound: Theory and industrial applications* (1998) Gordon and Breach, London.

5. L. Crum and K. Hynynen, Sound therapy in *Physics World* (1996) 28.
6. Special Edition covering First International Meeting on Sonochemistry, *Ultrasonics* (1987) *25* (1), January.
7. J. Thornycroft, and S. W. Barnaby, *Inst. C. E.*, (1895) *122*, 51.
8. K. S. Suslick, R. E. Cline, and D. A. Hammerton, *J. Am. Chem. Soc.* (1986) *108*, 5641.
9. T. G. Leighton, The principles of cavitation in *Ultrasound in Food Processing*, (1998) 151–82, ed. M. J. W. Povey and T. J. Mason, Thomson Science, London.
10. G. Cum, R. Gallo, A. Spadaro, and G. Galli, *J. Chem. Soc., Perkin Trans 11*, (1988) 376.
11. J. C. de Sousa-Barboza, C. Petrier, and J-L. Luche, *J. Org. Chem.* (1988) *53*, 1212.
12. A. Henglein, and M. J. Gutierrez, *J. Phys. Chem.* (1990) *94*, 5169.
13. P. Riesz, Free radical generation by ultrasound in aqueous solutions of volatile and non-volatile solutes, *Advances in Sonochemistry*, (1991) *2*, 23–64, ed. T. J. Mason, JAI Press, London, ISBN 1-55938-267-8.
14. K. S. Suslick, S-B. Choe, A. A. Chichowlas, and M. W. Grimstaff, *Nature* (1991) *353*, 414.
15. T. H. Hyeon, M. M. Fang, and K. S. Suslick, *J. Am. Chem. Soc.* (1996) *118*, 5492.
16. G. Price, The use of ultrasound for the controlled degradation of polymer solutions, in *Advances in Sonochemistry* (1990) **1**, 231–87, ed. T. J. Mason, JAI Press, London, ISBN 1-555938-178-7.
17. S. Folger, and D. Barnes, *Ind. Eng. Chem. Fundam.* (1968) *7*, 222.
18. E. C. Couppis and G. E. Klinzing, *I. Ch. E. J.* (1974) *20*, 485.
19. D. S. Kristol, H. Klotz, and R. C. Parker, *Tetrahedron Lett.* (1981) *22*, 907.
20. T. J. Mason, J. P. Lorimer, and B. P. Mistry, *Tetrahedron* (1985) *26*, 5201.
21. J. D. Sprich and G. S. Lewandos, *Inorg. Chim. Acta* (1982) *76*, 1241.
22. J. P. Lorimer, T. J. Mason, and D. Kershaw, *Colloid and Polymer Science* (1991) *269*, 392.
23. J. Lindley, P. J. Lorimer, and T. J. Mason, *Ultrasonics* (1986) *24*, 292.
24. K. S. Suslick and S. J. Doktyez, Effect of ultrasound on solids and surfaces, *Advances in Sonochemistry* (1990) **1**, 187–230, ed. T. J. Mason, JAI Press, London, ISBN 1-555938-178-7.
25. N. K. Goh, A. Teah, and L. S. Chia, *Ultrasonics Sonochemistry* (1994) *1*, S43.
26. R. S. Davidson, A. Safdar, J. D. Spencer, and D. W. Lewis, *Ultrasonics* (1987) *25*, 35.
27. K. S. Suslick, and R. E. Johnson, *J. Am. Chem. Soc.* (1984) *106*, 292.
28. D. J. Walton and S. S. Phull, Sonoelectrochemistry, in *Advances in Sonochemistry* (1996) *4*, 205–85, ed. T. J. Mason, JAI Press, London, ISBN 1-55938-793-9.

29. T. Ando, and T. Kimura, Ultrasound in solid supported chemical reactions, in *Advances in Sonochemistry* (1991) *2*, 211–58, ed. T. J. Mason, JAI Press, London, ISBN 1-55938-267-8.
30. C. L. Bianchi, R. Carli, S. Lanzani, D. Lorenzetti, G. Vergani, and V. Ragaini, *Ultrasonics Sonochemistry* (1984) *1*, S47.
31. F. Galton, *Inquiries into human faculty and development* (1883) MacMillan, London.
32. J. Curie, and P. Curie, *Compt. Rend.* (1880) *91*, 294.
33. A. A. Atchley, and L. A. Crum, Acoustic cavitation and bubble dynamics, in *Ultrasound its chemical physical and biological effects* (1998) ed. K. S. Suslick, VCH Publishers, Weinheim.
34. E. G. Janzen, *Acc. Chem. Res.* (1971) *4*, 31.
35. P. Riesz, T. Kondo, and C. M. Krishna, *Ultrasonics* (1990) *28*, 295.
36. R. W. Matthews, *Rad. Res.* (1980) *83*, 27.
37. M. Anbar, D. Meyerstein, and P. Neta, *J. Phys. Chem.* (1966) *70*, 2660.
38. J. R. McLean, and A. J. Mortimer, *Ultrasound Med. Biol.* (1988) *14*, 59.
39. A. J. Carmichael, M. M. Mossoba, P. Riesz, and C. L. Christman, *IEEE Trans. Ultra. Ferr. Freq. Cont.*, UFFC-33(2), 148, March 1986.
40. R. D. Corsaro, J. D. Klunder, and J. Jarzynski, *J. Acoust. Soc. Am.* (1980) *68*, 655.
41. P. Riesz, D. Berdahl, and C. L. Christman, *Envir. Health Persp.* (1985) *64*, 233.
42. T. J. Mason, and J. Berlan, Dosimetry in sonochemistry in *Advances in Sonochemistry* (1996) *4*, 1–73, ed. T. J. Mason, JAI Press, London, ISBN 1-55938-793-9.
43. J. P. Perkins, Power ultrasound in *Sonochemistry* (1990) 47–59, ed. T. J. Mason, Royal Society of Chemistry, London, ISBN 0-85186-293-4.
44. R. D. Corsaro, J. D. Klunder, and J. Jarzynski, *J. Acoust. Soc. Am.* (1980) *68*, 655.
45. A. Weissler, H. W. Cooper, and S. Snyder, *J. Acoust. Soc. Am.* (1948) *20*, 589.
46. N. Miller, *Trans. Faraday Soc.* (1950) *46*, 546.
47. J. A. La Verne and R. H. Schuler, *J. Phys. Chem.* (1987) *91*, 5770.
48. R. J. McLean and A. J. Mortimer, *Ultrasound in Medicine and Biology* (1988) *14*, 59.
49. T. J. Mason, J. P. Lorimer, D. M. Bates, and Y. Zhao, *Ultrasonics Sonochemistry* (1994) *1*, S91.
50. A. Kotronarou, G. Mills, and M. R. Hoffmann, *J. Phys. Chem.* (1991) *95*, 3630.
51. C. Petrier, A. Jeunet, J-L. Luche, and G. Reverdy, *J. Am. Chem. Soc.* (1992) *114*, 3148.

2 The ultrasonic cleaning bath

The simple ultrasonic cleaning bath is by far the most widely available and cheapest source of ultrasonic irradiation for the chemical laboratory. It is for this reason that so many sonochemists begin their studies using cleaning baths. Unfortunately, despite its convenience it is not a particularly powerful source and so sometimes researchers can be disappointed with results. Nevertheless a lot can be achieved even with such a simple piece of apparatus.

2.1 Bath construction

The construction of an ultrasonic cleaning bath is very simple—a laboratory model generally consists of a stainless steel tank of rectangular cross-section with either one or several transducers (depending on bath size) firmly attached underneath the flat base. Generally piezoelectric transducers are used and these are bonded with epoxy resin. Some tanks, particularly the larger sizes, also have some form of thermostatted heater. A few modern laboratory-scale models have adjustable power but these are in a minority (Fig. 2.1). The frequency and power of an ultrasonic bath depends upon the type and number of transducers used in its construction. Generally the ultrasonic power available in a bath using modern piezoelectric transducers is of quite low intensity of the order of 1–5 W cm^{-2} with an operating frequency of

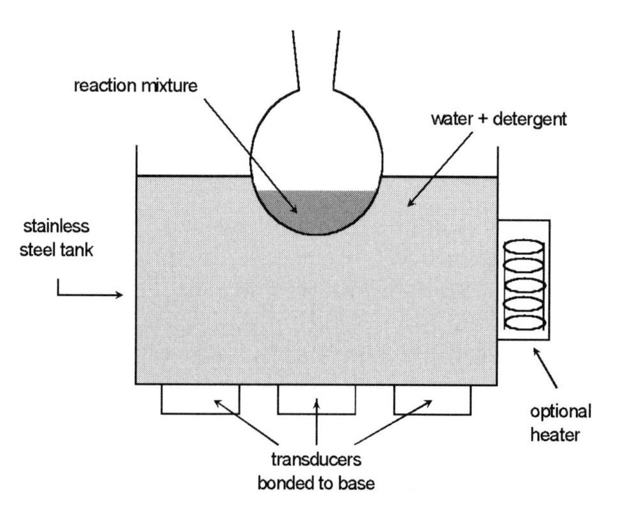

Figure 2.1 Ultrasonic cleaning bath for sonochemistry

approximately 40 kHz. There are a few variants on this design in that some baths operate at different (but fixed) frequencies. Some cleaning baths also operate with a 'frequency sweep' of a few kHz about a central frequency. The sweep is designed to avoid standing waves in the tank. In this way every part of an article for cleaning is thus subjected to the cavitation zone since the nodes in the waves alternate in height from the base.

The smallest baths suitable for sonochemistry are of about 1.5-litre capacity and a single transducer is sufficient to drive this with a power rating of around 50 W. Larger systems up to 50 000 litres are commercially available, and employ an array of transducers on the base to introduce uniform power density into the liquid. A real problem when trying to reproduce work reported in the literature is that ultrasonic baths operate at frequencies and powers dependent upon the transducers employed and with geometries which are specific to the particular manufacturer. This means that it is difficult to make direct comparisons between studies done in different makes (or models) of ultrasonic bath.

The liquid in the bath will normally be water containing a small quantity of surfactant. This is termed the coupling medium because it acts as the conduit for ultrasonic energy from the transducers in the base into the water and thence into any container dipped in the water. When water is used the upper limit for operational temperatures cannot exceed 100 °C. A change in medium will allow a greater range of temperature to be studied but this will also result in a change in energy transfer from transducer to the contents of the reaction vessel. Temperature control in baths around ambient temperatures is poor because the passage of power ultrasound through the bath liquid generates a small amount of heat. The system can be operated under thermostatic control at temperatures above that which is reached on thermal equilibration of the bath under its normal working conditions. It is also vitally important to record the temperature inside the reaction vessel during sonication since this generally tends to be a few degrees above that of the bath itself due to ultrasonic heating through the walls of the immersed reaction vessel. Temperature control is dealt with in detail in Section 2.3 below.

2.2 Choosing the correct type of ultrasonic bath

Not all commercially available ultrasonic cleaning baths are suitable for sonochemistry, and some baths, after extended usage, can lose power either through the denaturing of the piezoelectric transducer or a weakness in the transducer bonding to the base. The ideal bath must generate sufficient power to promote cavitation within the reaction vessel when it is immersed in the bath, i.e. enough power to penetrate the walls of the reaction vessel. Naturally the bath must also be large enough to accommodate the reaction vessel used.

If there is a bath available in the laboratory, it can easily be established whether it is suitable for sonochemistry by subjecting it to the foil test. The bath should first be filled with water containing some 1 or 2 per cent surfactant—this reduces the surface tension of the water and permits much more even cavitation in the bath water. Indeed when the bath contains just tap water a distinct change in 'pitch' can be detected as the detergent is added (see Section 2.5). A piece of ordinary aluminium kitchen foil about six inches square should be dipped into the bath (it is best to avoid immersing fingers into the sonicated bath liquid). After 30 seconds remove the foil and if it is liberally perforated then the bath has passed its test and is powerful enough for sonochemistry.

2.3 Setting up an ultrasonic cleaning bath for sonochemistry

Although it is possible to use the bath itself as a reaction vessel this is seldom done because of the problems involved in corrosion of the bath walls and containment of any evolved vapours and gases. Normal usage therefore involves the immersion of standard glass reaction vessels into the bath [1]. This is of great importance, since conventional glass apparatus can simply be transferred into the bath—this means that an inert atmosphere or a static pressure can be readily achieved and maintained throughout a sonochemical reaction which involves an ultrasonic bath.

Positioning the reaction vessel

It is a common mistake in sonochemistry to assume that all that is required to achieve significant results is that the reaction vessel should simply be immersed in the water and the powers switched on. Consider the way in which the bath has been constructed. The ultrasonic transducer(s) will, most likely, be attached to the underside of the metallic base of the bath and the most intense power will be found directly above a transducer. However, the ultrasonic power in the bath liquid will not be uniform with distance from the base. This is because ultrasound, like any sound, passes through water in the form of a wave, and the wave will have positions of maximum amplitude at multiples of the half-wavelength of sound (λ) in the medium. These distances may be calculated using the simple eqn (2.1).

$$v = f\lambda \tag{2.1}$$

The velocity of sound (v) through water is approximately 1500 m s^{-1} so that for a transducer operating at a frequency (f) of 40 kHz the wavelength of the ultrasound in water will be 3.75 cm. We may expect the maximum effect to occur at vertical intervals of 1.8 cm. This series of horizontal regions of high intensity can be located by sonicating a large piece of aluminium foil placed vertically in the bath. A number of distinct horizontal series of perforations appear after some 30 seconds immersion. Towards the base of the bath, close to the emitting surface, there is confusion in the standing wave pattern. In this nearfield region, the standing waves are much closer together (Fig. 2.2). For

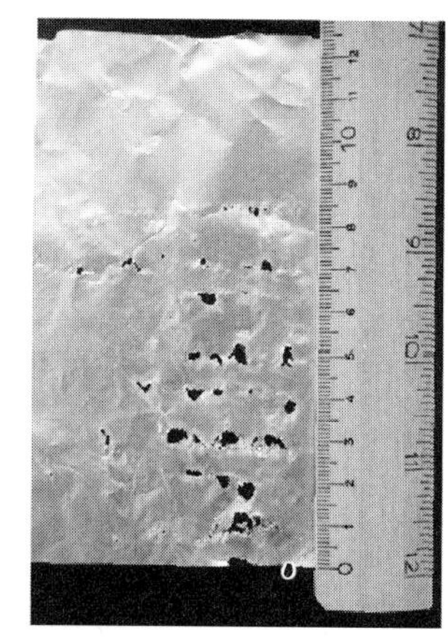

Figure 2.2 Effect of cavitation of aluminium foil immersed in an ultrasonic cleaning bath (40kHz)

larger baths where measurements can be taken at greater distances from the emitter the standing wave pattern reverts to half-wavelengths (see Fig. 2.5).

Using a series of foil sheets the power regions of the bath can be quite accurately identified because the maximum perforations occur at maximum intensity.

Naturally the reaction vessel should be located at the point where the maximum sonochemical effect can be achieved. Observation of the surface of the reaction solution during vertical adjustment of vessel depth will show the optimum position by the point at which maximum surface disturbance occurs. For a bath with a single transducer on the base this is a simple task since the reaction vessel must be located vertically above the single transducer. For a bath with multiple transducers the foil 'mapping' technique can be used to locate the regions of most intense power for horizontal positioning before the depth is established. Another approach to the determination of the way in which the 'sonic energy' is distributed within the bath is the use of thermistor probes [2]. Pugin has shown that the distribution of ultrasonic energy is not at all homogeneous by measuring the local temperature at different places in the sonicated media [3]. Maxima and minima appear, depending on the shape of the vessel, the liquid height, and the input power. Once the correct position for the reaction vessel has been identified it is a simple matter to arrange stands and clamps in order to locate it in the required position. Essentially this will act as a template for further reactions, since repeated use of the same vessel in the same location in the bath should give reasonably reproducible results.

Temperature control

All ultrasonic baths warm up under the influence of the transducers in the base, and this creates certain problems in terms of both the effectiveness and reproducibility of the sonochemical results obtained. The simplest solution to the problem is to determine the 'equilibrium' temperature of the bath (i.e. the maximum temperature which the water reaches and maintains when operating continuously under ambient conditions) and perform most reactions under these conditions. For small baths this is about 45 °C and so, before starting any experiments, it is advisable to fill it with hot water so that equilibrium temperature is attained in the shortest time possible. It is possible to thermostat the bath above this temperature using a heater. Such heaters are normally integrated into cleaning baths attached to the outside of the tank walls.

Three solutions are available to the problem of operating at room temperature or below:

(i) operate for very short periods with cold water in the tank (the temperature can be assumed to remain essentially constant for short bursts of sonication);

(ii) circulate thermostatted cooling water through the tank; or

(iii) add ice.

If the bath is equipped with a cooling coil or ice is used, it must be remembered that solids and inserts will disturb the normal propagation of ultrasound through the bath and so alter its characteristics. Whatever method is chosen for temperature control, it must be emphasised that it is the temperature inside the reaction vessel which must be monitored as this is often a few degrees above that of the bulk bath liquid itself.

2.4 The design of the reaction vessel

Once the bath has been chosen a reaction vessel of the correct design must be chosen. For normal chemical reactions, particularly those involving heat, round-bottomed flasks are utilised, but for sonochemistry in an ultrasonic bath the preferred vessel should be one with a flat bottom, e.g. a conical flask. The reason for this is that the energy is radiating vertically as sound waves from the base of the bath and this energy has to be transferred through the glass walls of the vessel into the reaction itself. The energy transference is much more effective when the sound impinges directly on the flat base of a conical flask rather than hitting the underside of a spherical container at an angle, since when this happens some energy will be reflected away. Another important consideration when using baths to perform sonochemical reactions is that it may be necessary to stir the mixture mechanically to achieve the maximum effect of the ultrasonic irradiation. This is particularly important when using solid – liquid mixtures where the solid is neither dispersed nor agitated throughout the reaction by sonication alone and simply sits on the base of the vessel where it is only partially available for reaction. The reasons why additional stirring is so important in such cases is that:

(a) it ensures the reactant powder is exposed as fully as possible to the reaction medium during sonication; and

(b) with stirring there will be no layer of powder on the base of the vessel which could produce considerable attenuation of the incident sound energy.

A practical comparison of vessels for use with ultrasonic baths

A round-bottomed flask was compared with two different sized conical flasks to illustrate the most appropriate vessel shape for use with an ultrasonic bath [4]. A small single transducer bath was used in this exercise, a Kerry Pulsatron 60. This bath operates at 38 kHz, is rated at 50 Watts, and has internal dimensions of 14×15 cm and a depth of 10 cm. In each experiment the flask was positioned for maximum sonochemical effect (see above).

Homogeneous model

As a model for a homogeneous reaction, the liberation of iodine by sonic irradiation of 4 per cent aqueous potassium iodide (50 cm^3, containing 2.5 cm^3 carbon tetrachloride) was chosen. The reaction was monitored by periodically withdrawing aliquots and measuring any increase in the liberation of iodine by increases in its absorption spectrum at 351 nm. The results obtained at 30 °C (Fig. 2.3) illustrate the great importance of both the source

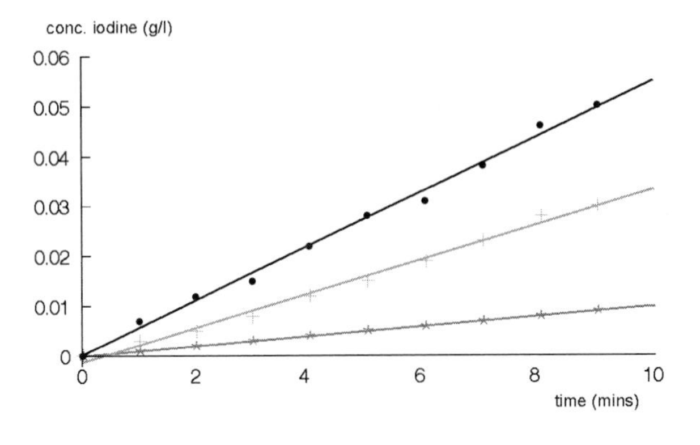

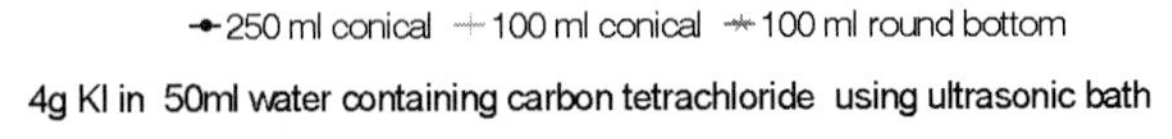

Figure 2.3 Dependence of iodine liberation on vessel shape

of ultrasound and the type of reaction vessel chosen for sonochemical studies.

For a reaction volume of 50 cm^3 a larger effect is obtained in a 250 cm^3 rather than a 100 cm^3 conical flask. Both vessels present the preferred configuration for transfer of energy from the bath to the reaction media, i.e. a flat glass base at right angles to the propagating sound wave. The results suggest that the greater base area of the 250 cm^3 flask allows more power input to the system—and hence a more effective sonochemical reaction. The use of a 100 cm^3 round-bottomed flask clearly results in a reaction which is inferior to either of these.

Heterogeneous model

The heterogeneous system chosen was the reduction in particle size of commercial potassium carbonate (1 g) by sonication in dimethylformamide (50 cm^3). Particle sizes were monitored using a Gallai instrument and the lower limit of particle size reduction was arbitrarily taken to be the point at which the mixture became too cloudy for the instrument to function properly (Fig. 2.4).

Particle fragmentation using a 100 cm^3 vessel make it clear that to achieve optimum results mechanical stirring of the heterogeneous mixture is vital. The trend in results for particle size reduction in differently shaped vessels is the same as for the homogeneous reaction in that the flat-bottomed conical was more effective than the round-bottom flask. From these results it can be concluded that the flat bottom of a conical flask is the most effective shape for use in an ultrasonic bath, but that the best results are obtained with vessels where there is a large base area with respect to volume. Clearly stirring is essential for solid/liquid heterogeneous reactions.

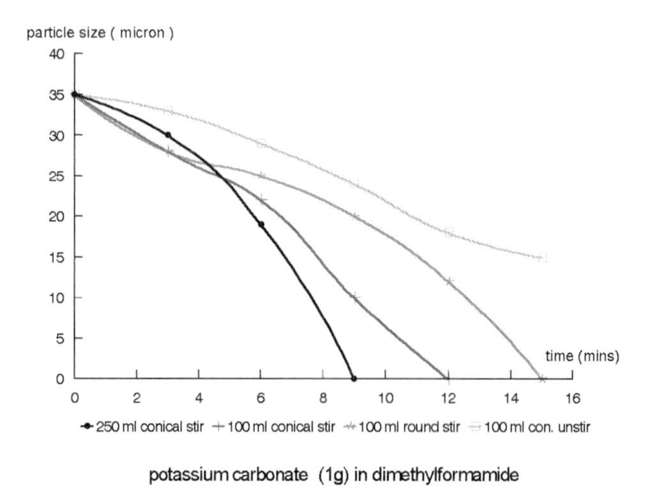

Figure 2.4 Dependence of particle size reduction on vessel shape

2.5 Laboratory demonstrations of the physical effects of ultrasound

The short experiments described below have been developed as demonstrations which have been used in lectures and seminars on sonochemistry. They are all of short duration and mainly show aspects of the mechanical effects of power ultrasound. In these examples the bath can be used from cold—no special heating or thermostatting is required.

The effect of added detergent

The bath was filled with pure distilled water and switched on. An audible sound is emitted from the tank due to metal vibrations and the subharmonic content of the ultrasonic source. After a short period during which the water is degassed the sound settles and becomes irregular in volume, the surface is often subject to erratic disturbance during this period. The addition of about 1–2 per cent detergent changes the note of the sound and causes a more even and enhanced disturbance of the surface. The reasons for this have been explained above, i.e. the detergent has produced much better coupling between the ultrasonically reverberating base of the tank and the liquid medium contained in it. Hence the bath has become a much more reproducible source of ultrasound.

Ultrasonic degassing

The cavitational effects which are the basis of sonochemical action are also the reason for the extremely effective use of ultrasound to degas liquids. Once the cavitation bubbles have been formed they fill with any dissolved gas. Such bubbles are not easily collapsed in the compression cycle of the wave due to the fact that they contain gas, and they will continue to grow on further rarefaction cycles, filling with more gas and eventually floating to the surface. Since the rarefaction cycles are taking place extremely rapidly (around 40 000 times per second using an ultrasonic bath), the bubbles grow so quickly that degassing appears to occur almost instantaneously. The simplest way to demonstrate the effect is to place a flask of sparkling water in an ultrasonic bath. The immediate result is that the liquid is caused to effervesce violently, and the effervescence drops immediately as the flask is removed only to be resuscitated on replacement in the bath.

An alternative approach is to produce the gasified liquid from granular zinc metal in 4 M hydrochloric acid (125 cm^3) in a 250 cm^3 conical flask. After the effervescence has started (about 30 seconds) the liquid becomes opaque with liberated hydrogen. When the flask is touched onto the surface of the water in the ultrasonic bath the cloudiness is cleared as a front which rises from the base. The process can be repeated as long as the zinc remains. It is also worth noting that the production of the gas at the metal surface also changes, from the very large number of tiny bubbles which generate cloudiness to larger bubbles in the ultrasonic field which rapidly rise to the surface leaving the liquid clear. Ultrasonic degassing has found applications in many areas. In its simplest form it is used to degas solvents for use in HPLC. The electroplating of metals invariably involves gas generation at one or other of the electrodes. The physical presence of the gas acts as a barrier to efficient passage of current (discharge of ions), ultrasound can improve electroplating and indeed general electrochemical processes by removing this gas 'barrier' [5]. A somewhat different development involves use of ultrasonic equipment for the removal of volatile constituents in crude oil fractions, i.e. the 'degassing' of low-boiling material such as methane and ethane [6].

Reactions involving metal surfaces

There are two types of reaction involving metals, one in which the metal is a reagent and is consumed in the process, and a second in which the metal functions as a catalyst. Metals which are ultrasonically irradiated in a solvent suffer considerable surface damage. This effect is caused by two possible processes:

(i) the implosion of cavitation bubbles formed from seed nuclei on the surface; or

(ii) microstreaming of a jet of solvent onto the surface when a cavitation bubble collapses close to it.

The result of this pitting is to expose new, clean metal surface to the reagents and also to increase the effective surface area available for reaction. The perforation of foil (see above) is a perfect example of this effect. A clean piece of aluminium kitchen foil (15 × 5 cm) is placed vertically in the sonicated bath liquid and kept still for 30 seconds. On removal the foil shows perforations in lines parallel with the surface of the liquid. The horizontal rows corresponding to the areas of maximum cavitational effect at half-wavelength distances vertically through the bath liquid.

Reactions involving powders

One of the easiest processing operations in sonochemistry is the deagglomeration of powders. This can be demonstrated using blackboard chalk as the example of the powder. A few pieces of chalk are placed in water (150 cm^3) in a 250 cm^3 conical flask. When the flask is dipped into the ultrasonic bath, clouds of chalk can be seen emanating from the surface of the solid. Very quickly the liquid becomes cloudy illustrating the way in which ultrasound can erode surfaces and break down agglomerated particles to a much smaller size.

Emulsion reactions

Ultrasound is known to generate extremely fine emulsions from mixtures of immiscible liquids. One of the main consequences of these emulsions is the dramatic increase in the interfacial contact area between the liquids, i.e. an increase in the region over which any reaction between species dissolved in the different liquids can take place. This can be shown by placing some water (100 cm^3) and methylcyclohexane (50 cm^3) in a 250 cm^3 conical flask and dipping the flask into an ultrasonic bath. By carefully adjusting the depth of immersion the clearly defined phase boundary between the two immiscible liquids is seen to become agitated. On closer inspection this disturbance can be seen as a large number of tiny 'explosions' at the interface effectively sending small jets of liquid from one phase into the other. Soon a cloudiness appears in the interfacial region and within a minute or so substantial emulsification will have been caused. The demonstration is made even more dramatic if water soluble dye is added to make the aqueous phase brightly coloured.

The separation of particles in an acoustic field

This experiment uses a suspension of copper bronze. Copper bronze is finely divided copper with particle sizes between 50 and 100 μm (1 g in 100 cm^3 water containing 1 per cent detergent). The mixture is shaken and poured into a 100 cm^3 measuring cylinder to give a uniformly bronze coloured column of liquid. When the base of the cylinder is placed in the ultrasonic bath liquid the uniform suspension changes to a set of horizontal bright metallic lines at separated distances of the half-wavelength of the ultrasound in the solution (which is around 1.8 cm at 40 kHz) (Fig. 2.5). These striations are produced because the copper bronze particles collect at the nodes of the sound waves in the water. This demonstrates a potentially useful application of ultrasound as a method of manipulating small particles in a liquid medium [7].

Homogeneous generation of iodine

Power ultrasound is capable of generating hydrogen peroxide from the fragmentation of water (see Chapter 1). The peroxide will oxidise any iodide ions in solution to produce iodine. Since the power of an ultrasonic bath is not very great it is normally not possible to see the pale yellow colour generated by the small amount of iodine produced by the sonication of aqueous KI. It is, however, possible to detect the low level of iodine liberated through its reac-

Figure 2.5 Effect of ultrasound on copper bronze suspended in water

tion with starch to produce the characteristic blue colour. Even this will take a considerable time to generate using a 40 kHz bath. The effect can be enhanced by the addition of a few drops of carbon tetrachloride to the KI solution. Place 4 per cent aqueous KI solution (100 cm^3) in a 250 cm^3 conical flask and add a few drops of CCl_4 followed by soluble starch indicator (2 cm^3). The flask is then immersed in the ultrasonic bath to a depth where the clear solution suffers maximum disturbance. Within one minute the blue starch/iodine colour will start to appear and this will deepen as the reaction proceeds.

Sonochemically enhanced chemiluminescence

Luminol degrades with the liberation of light in the presence of hydroxy radicals and hydrogen peroxide (Scheme 2.1). This reaction can be intensified if the radical and peroxide generation is concentrated into a small region. Such conditions occur at the interface between carbon tetrachloride and water during sonication.

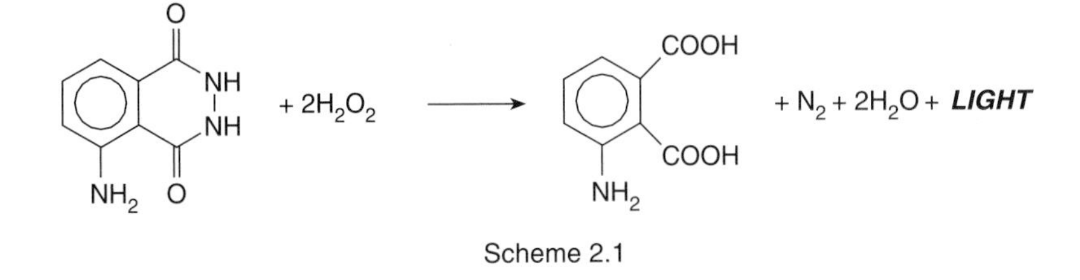

Scheme 2.1

This demonstration requires that luminol (0.04 g) and sodium carbonate (1 g) are dissolved in distilled water (100 cm^3) in a 250 cm^3 Erlenmeyer flask. To this solution a few drops each of hydrogen peroxide (6 per cent or 20 volume) and carbon tetrachloride are added. In darkness the flask should be dipped into the ultrasonic bath. A series of bright sparks of light is emitted from the carbon tetrachloride/water interface as it is agitated by the sonic waves. These sparks stop when the flask is removed from the tank but are regenerated on dipping it in again. This effect can be repeated over a considerable period of time.

2.6 Advantages and disadvantages of using an ultrasonic bath for sonochemistry

Despite the fact that the cleaning bath is the piece of ultrasonic equipment most widely used by chemists it is not necessarily the most effective. The advantages and disadvantages are summarised below.

Advantages of using an ultrasonic bath in sonochemistry

1. It is the most widely available laboratory source of ultrasonic irradiation.

2. A small cleaning bath is inexpensive.

3. The acoustic field is fairly evenly distributed throughout the bath liquid.

4. There is no special adaptation of chemical apparatus required, conventional glassware can be used, which means that the addition of chemicals, an inert atmosphere, and the use of reduced and elevated pressures are all easily attained.

5. If the cleaning bath itself is used as the reaction vessel, larger batch treatment under greater irradiation power can be achieved. Despite the advantages gained from the use of such a simple piece of apparatus, there are a number of considerations which should be borne in mind when using this method of energy input.

Disadvantages of using an ultrasonic bath in sonochemistry

1. The amount of power dissipated into the reaction from the bath is not very large—usually less than $5\,W\,cm^{-2}$.

2. The energy input should be assessed for each system investigated because power will depend on the size of the bath, the reaction vessel type (and thickness of its walls), and the position of the reaction vessel in the bath.

3. The ultrasonic frequency at which a cleaning bath operates is not universally the same. This may well affect results particularly when attempting to reproduce those reported in the literature. Most manufacturers use frequencies around 40 kHz, but 20 kHz baths are available and higher frequencies may be used in some specialist baths.

4. Temperature control is not easy since most cleaning baths warm up slowly during operation. This may lead to inconsistent results when working around room temperature or below. Solutions to this problem have been given (see Section 2.3).

5. A few manufacturers produce cleaning baths with adjustable power, but the majority of cleaning baths do not have any such control. It is possible to place a rheostat between the mains and the bath which will permit simple power control.

For all commercial ultrasonic cleaning baths it is assumed that the sonochemical reaction will be performed in a vessel immersed in the sonicated liquid. It is possible to use the whole bath as a batch reactor but this can only apply where non-corrosive and non-volatile reagents are used. Very few cleaning baths have vapour-tight lids, which means that it is very difficult to use them for reactions which require an inert atmosphere or a reflux condenser.

It is, however, clear from the variety and large number of experiments reported in the literature that despite its drawbacks the simple ultrasonic cleaning bath is extremely useful for sonochemistry.

2.7 Other types of equipment related to cleaning baths

Once a laboratory cleaning bath has been shown to be of sufficient energy to promote a sonochemical reaction, the next step may be to perform a more rigorous investigation. In this way the process can be optimised or adapted to a larger scale. There are a variety of pieces of equipment which are available for this purpose.

The submersible transducer

For many years the manufacturers of ultrasonic cleaners have offered the submersible transducer as a convenient and versatile method of converting any vessel into an ultrasonic cleaning bath. The submersible consists of a number of transducers which have been sealed into a stainless steel box (Fig. 2.6). These are bonded to the inner face of the box which then becomes the radiating face when the submersible is immersed into a liquid. Normally the submersible is equipped with hooks to hang over the edge of the bath and a flexible (water tight) lead for connection to the generator.

The use of a submersible system to convert a tank into a sonicated bath affords several advantages over the cleaning bath.

1. A container of almost any size and shape can be converted into an ultrasonic bath using a submersible. These are available off the shelf, or can be constructed, in a range of sizes.

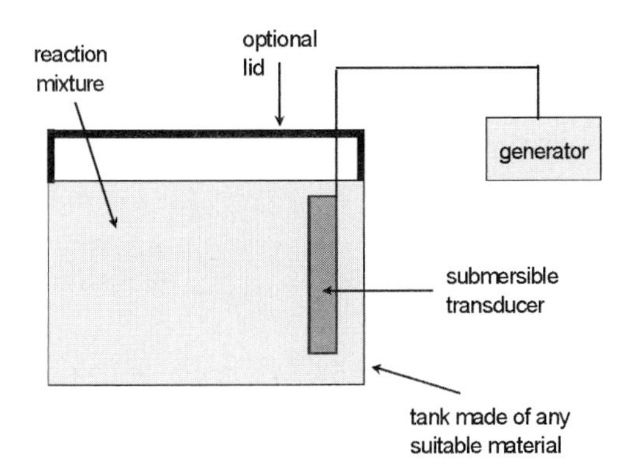

Figure 2.6 Submersible transducer system

2. The submersible transducer can be dipped directly into a chemical reaction thereby introducing power directly into a system, the equivalent of using the cleaning bath itself as the reactor. Of course this advantage will only be of use if the system does not chemically react with the transducer housing.

3. The position of the submersible can be adjusted to give maximum sonochemical effect in the particular vessel used.

4. The vessel to be used can be quite large since any number of submersibles can be used.

The most likely sonochemical applications for submersible transducers are in the batch treatment of aqueous or non-corrosive liquid systems.

The cup-horn

If the general utility of an ultrasonic bath could be combined with controllable power and temperature control of a probe, a useful compromise would be available. Such a compromise system is the cup-horn device (Fig. 2.7), which can be considered to be a small but powerful ultrasonic bath. In effect, it uses a specially designed inverted horn to irradiate a small cup of liquid (normally water) into which the reaction vessel is dipped. The horn, with a hole drilled through its centre, allows the liquid in the cup to be part of a circulating coolant system which assists in maintaining temperature stability. Since the irradiation source is a modified sonic horn this system has one major advantage over the cleaning bath—its power can be controlled.

The cup-horn was originally designed for use in biological cell disruption, which requires higher power than the normal cleaning bath. The other major advantage of this source over the probe system is that since the cellular material is in a vessel immersed in the sonicated fluid, and thus not in direct contact with the irradiating surface, it avoids contamination by fragments of

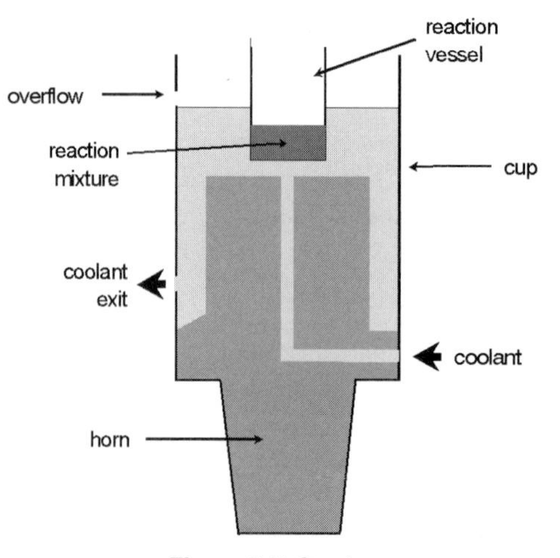

Figure 2.7 Cup-horn

metal which might be eroded from a directly immersed probe system. The main difficulty with the cup-horn is that it is quite small and this limits the size of the reaction vessel which can be used.

High-frequency ultrasonic baths specially designed for sonochemistry

The ultrasonic bath described below is the type developed by Arnim Henglein, who began his studies of the chemical effects of ultrasound as a student studying for a masters degree over 50 years ago [8]. Today most chemists use commercial ultrasonic equipment which operate in the frequency range 20 to 40 kHz, but Henglein often conducted experiments at a much higher frequency (1 MHz). A vessel which is capable of high insonation intensities at 1 MHz is shown in Fig. 2.8. The glass bottom is a half-wavelength thick and the vessel is placed directly on the aluminium metal flange which carries the transducer (a thin film of water between the vessel and the flange assures efficient acoustic coupling).

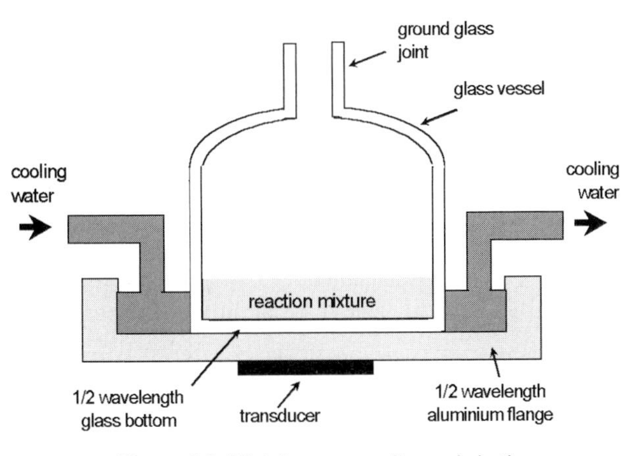

Figure 2.8 High frequency ultrasonic bath

In recent years, there have been a variety of higher frequency baths developed for sonochemistry. The drive for these developments has been the discovery that at higher irradiation frequencies (e.g. 500 kHz) the characteristics of cavitation bubble generation and collapse are different from those which occur at 20 and 40 kHz (see Section 1.4). Radicals also are produced more efficiently [9]. In essence most of these baths are constructed in a similar way with a planar, usually titanium, emitting face sealed into the base of a glass vessel with external cooling (Fig. 2.9).

2.8 The power of equipment related to cleaning baths

In any apparatus related to cleaning bath systems the cavitation bubbles are formed as separate entities throughout the sonicated liquid medium. This

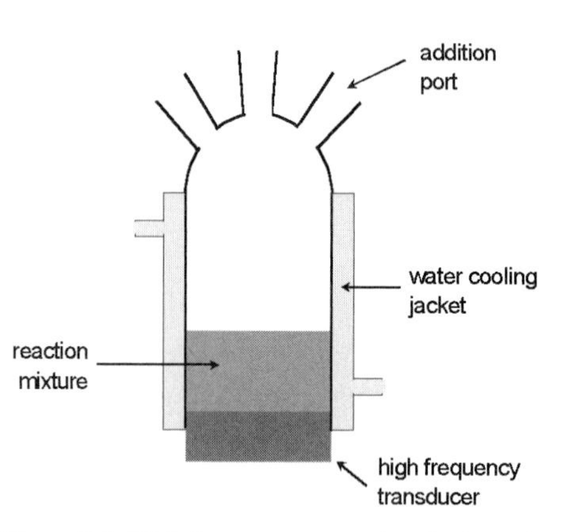

Figure 2.9 High frequency sonochemical reactor

contrasts with the streaming and coalescence of bubbles which occur in the more powerful sonicators, e.g. probe systems. It is therefore difficult to compare the results of experiments carried out with bath and probe type sonication equipment. There is no doubt, however, that cleaning baths provide sufficient power to identify reactions which are susceptible to sonochemical enhancement, and that they are certainly the most convenient to use under laboratory conditions.

References

1. T. J. Mason, J. P. Lorimer, and J. Moorehouse, *Education in Chemistr* (1989) *26*, 13.
2. C. J. Martin and A. N. R. Law, *Ultrasonics* (1983) *21*, 85.
3. B. Pugin, *Ultrasonics* (1987) *25*, 49.
4. T. J. Mason, J. P. Lorimer, F. Cuesta, and L. Paniwnyk, *Ultrasonics International 89*, Conference Proceedings (1989) 1253.
5. T. J. Mason, J. P. Lorimer, and D. J. Walton, *Ultrasonics* (1990) *28*, 333.
6. N. Senapati, Ultrasound in chemical processing in *Advances in Sonochemistry* (1991) *2*, 187–210, ed. T. J. Mason, JAI Press, London, ISBN 1-55938-267-8.
7. C. J. Schram, The manipulation of particles in an ultrasonic field in *Advances in Sonochemistry* (1991) *2*, 293–322, ed. T. J. Mason, JAI Press, London, ISBN 1-55938-267-8.
8. A. Henglein, Sonochemistry—a historical perspective in *Advances in Sonochemistry* (1992) *3*, 17–84, ed. T. J. Mason, JAI Press, London, ISBN 1-55938-467-X.
9. C. Petrier, A. Jeunet, J-L. Luche, and G. Reverdy, *J. Am. Chem. Soc.* (1992) *114* 3148.

3 The ultrasonic probe

Despite its convenience for use, the major disadvantages of the ultrasonic cleaning bath as a general source of ultrasound for sonochemistry are a low maximum power and the absence of energy control. One solution to both of these problems has been found in a piece of equipment originally developed for plastic welding and biological cell disruption—the ultrasonic probe. In the laboratory, it is possible to introduce ultrasonic energy into a reaction using a probe system up to some 100 times greater than can be achieved using a cleaning bath. On a larger scale the energies can be much greater than this. Modern probe systems are generally based on piezoelectric transducers and are all of a similar construction.

3.1 Ultrasonic probe construction

In order to increase the amount of ultrasonic power available to a reaction it is desirable to introduce the energy directly into the system, rather than rely on its transfer through the water of a tank and then the reaction vessel walls. The simplest method of achieving this would be to have the face of an ultrasonically vibrating transducer in contact with the reaction medium. The maximum amount of power (vibrational amplitude) generated by a transducer assembly is limited by the properties of the material from which it is constructed. A piezoceramic will fail if too much electrical power is fed into it (see Section 1.7). It is possible to amplify the vibrational amplitude generated at the transducer by attaching to it a specially designed length of metal rod. This rod extension is termed a sonic horn or velocity transformer, and it not only magnifies the acoustic energy available but also allows the transducer to be kept clear of the chemical reaction since only the tip of the rod needs to be immersed in the liquid. It is the complete assembly of transducer plus horn which is referred to as an ultrasonic probe system. The overall construction of a simple laboratory probe system is shown in Fig. 3.1.

The generator

This is the source of alternating electrical frequency (normally 20 kHz) which applies the transducer. The basic controls on the generator are given below.

Tuning

For optimum performance a probe system should be 'tuned' before it is used. Tuning is a process whereby the complete probe assembly is brought into resonance with the transducer—this is the point where the minimum current

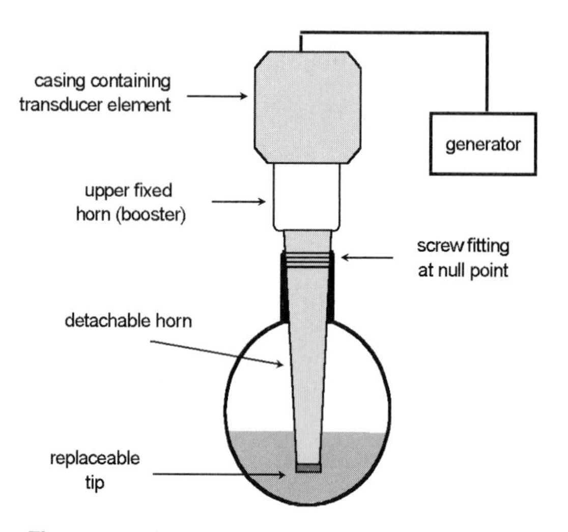

Figure 3.1 Ultrasonic probe system for sonochemistry

is drawn during operation. This may well be at a frequency fractionally different from 20 kHz. Thus, if a horn is shortened by tip erosion, then the resonance frequency will change. Eventually the limits of adjustment of the generator and transducer will be reached and tuning will no longer be possible—when this occurs the only solution is to replace the horn. Early types of generator were only able to cope electronically with small variations in tuning frequency. It was recommended that the initial tuning of such systems was carried out in air when the system is not 'under load'. The load refers to the extra power required to drive the probe system when it is immersed in a liquid medium, so the generator must also be able to provide the extra power to the transducer, and this forms the basis of one method of measuring the acoustic power input to a reaction (Section 1.9).

Sometimes a system which has been tuned in air will not perform satisfactorily when immersed in the reaction mixture. This is particularly noticeable when operating at low powers with viscous liquids. If the reaction mixture changes in physical characteristics during sonication—through a change from starting materials to products or warming—the power required to maintain optimum resonance changes, so that re-tuning may be necessary. The modern type of generator can tolerate a much wider range of tuning frequencies. Such systems are tuned when immersed in the reaction, and various types of electrical feedback can be incorporated in the design to give continuous adjustment of the tuning as the reaction proceeds. The slight change in the overall resonance frequency which results from changes in a liquid load are discussed below (Section 3.4).

Power

The power control on a simple generator simply allows the energy delivered to the transducer to be varied—normally on an arbitrary 0 to 100 per cent scale.

The upper limit prevents overload and fracture of the piezoceramic element, but the scale is not a direct reading of the energy entering the reaction.

Pulse

This was originally incorporated into probe systems to prevent the build up of reaction temperature during continuous operation. The facility consists of a timer attached to the amplifier which switches the power to the probe on and off repeatedly. The 'off' time allows the chemical system to cool between the pulses of sonication. The 'on' time is represented as a fraction of the total time involved in the total time of the cycle, which is generally set at 1 second. Thus 100 per cent is continuous sonication, while 25 per cent represents 0.25-second bursts of power interspersed with 0.75-second 'off' time. This type of pulse operation should not be confused with 'duty cycle' as used in medical ultrasound. In medical parlance, the duty cycle refers to the on/off ratio for scanning which involves the emission of pulses of extremely short duration (of the order of 10^{-5} s) involving only a few cycles of ultrasound in the megahertz range. It is in the longer off period that the echoes are detected. By comparison a pulse cycle of 50 per cent in a sonochemistry probe system corresponds to a 0.5-second 'pulse' which contains 10 000 cycles at 20 kHz.

The transducer element

The sandwich transducer element is protected by a casing (the transducer housing) which is perforated to allow cooling and may sometimes contain a small fan to prevent overheating. The shape and dimensions of the transducer assembly are dependent on its working frequency. Generally a 20 kHz assembly is twice the length and width of that of a 40 kHz. When a magnetostrictive transducer is used the element will also be protected in a casing, but in this case, the cooling is rather more vital and this is normally achieved using a circulating fluid.

The upper (fixed) horn or booster

The role of the booster is to adapt the vibration from the transducer so that it can be passed through the detachable horn at a working amplitude, which is introduced into the reaction medium. For the majority of laboratory probe systems this is the part of the system to which other horns of different configuration (see below) may be attached. The booster is a solid piece of titanium alloy which is accurately machined to provide a part of the overall resonating system. The end face of the fixed horn is normally about 1.2 cm in diameter, vibrating at a maximum in amplitude. It is important to recognise that with the (detachable) horn in place the whole probe system resonates, with a maximum vibration now occurring at the tip which is dipped into the reaction medium. With the whole probe unit in resonance there will be points of zero vibrational motion (null points) in the system and one of these will be on the booster. This is where a screw thread may be located for the attachment of optional ancillary equipment, e.g. a flow cell or a PTFE sleeve (see below).

The detachable horn

An important part of the whole probe assembly is the detachable horn, which allows the vibration of the fixed horn to be transmitted through a further length of metal which can be used to magnify the power delivered to a system. Most modern probe systems are designed to operate with a range of detachable horns, each offering a different magnification of power. Horn design is in fact a very important aspect of ultrasonic engineering. The material used for acoustic horns should have high dynamic fatigue strength, low acoustic loss, resistance to cavitation erosion, and also be chemically inert. The most suitable material by far for this purpose is titanium alloy.

The cavitation which is the source of chemical activation is also the source of a common problem with probe systems—tip erosion—which occurs despite the fact that most horns are made of titanium alloy. One major side effect of erosion is that there will be a physical shortening of the horn below a half-wavelength which will cause a loss of efficiency. The latter problem is alleviated by the use of replaceable screw-on tips for the 1.2 cm diameter horn in the form of titanium studs, this eliminates the need for a costly replacement of the whole section.

There are several engineering factors which are important in the design of the different types of detachable horn. The first of which is its length. The wavelength of ultrasound in a material is determined by both the type of material and the frequency of the sound wave. In the case of the type of titanium alloy used for horns, the wavelength for 20 kHz sound is about 26 cm, and this defines absolutely the longitudinal dimension of cylindrical horns made of titanium alloy. The minimum length of such a titanium horn is a half-wavelength, i.e. 13 cm, which will give an exact mirror of the vibrational amplitude supplied at one end to the other. If the distance between the upper fixed probe assembly and the sample being processed needs to be increased, this can be done using a horn cut to a multiple of half-wavelengths. This can also be achieved by screwing one half-wavelength horn into another thereby building up the overall length. This will not change the vibrational amplitude.

The shape of the horn controls the way in which it can be used to transmit acoustic energy from the booster to the reaction mixture (Fig. 3.2). The cylinder will not affect the vibrational amplitude, but if the end diameter of a horn is smaller than the area attached to the driving transducer element, there will be a magnification of the ultrasonic vibration (i.e. power). Consider ultrasound entering at the booster end of the horn (driven face D) at say 5 μm peak to peak at a cross-section of 3 cm diameter. All of this energy must be dissipated at the other end of the horn. If that end is of smaller area, say 1 cm diameter (emitting face d), then the vibrational amplitude must be greater in order to conserve energy. This amplitude gain is defined by a specific ratio of the diameter of the two faces of the horn. For a linear or exponential tapered horn, this ratio is D/d giving an emitting amplitude of 15 μm. For a stepped horn, however, the ratio is $(D/d)^2$ giving 45 μm. Thus, horn magnification is always higher for the stepped horn.

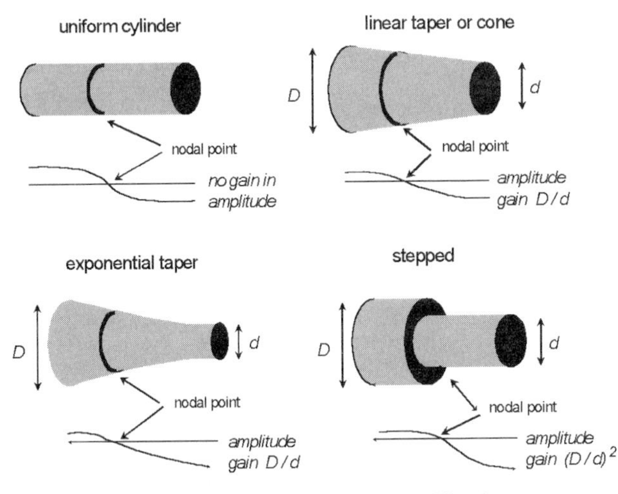

Figure 3.2 Horn shapes and amplification

In order to avoid material damage by internal stress, the amplification factor cannot be too high, for the linear taper it is normally restricted to four. The exponential taper offers higher magnification factors than the linear taper. Its shape makes it more difficult to manufacture, but the small diameter of the working end (3 mm or less) and its length make it particularly suited to micro-applications. The magnification factor for a simple 'stepped' design is significantly larger than the tapered cylinder and can be up to 16.

Probe systems are undoubtedly the most efficient method of transmitting ultrasonic energy into a reaction. Although so much extra power is available using the probe system, it is more expensive than the bath and it is significantly less convenient in use especially in terms of the glassware required. Special seals will be needed if the horn is to be used in any reactions which involve reflux, inert atmospheres, or the use of pressures above (or below) ambient.

3.2 Choosing a probe system for sonochemistry

For the probe system, depending on the design of horn, a large maximum power density can be achieved at the radiating tip. This can be of the order several hundred W cm^{-2}.

Conventional probe systems

Until quite recently the choice of commercial probe sonicators was somewhat limited. The only frequency available was 20 kHz and the peripheral equipment was also similar. Essentially this still remains the case, and although there may seem to be a large variety of names for the product, these tend to be the names attached by suppliers rather than those of the very few manufacturers. Nevertheless such systems are ideal for use as laboratory-scale sonicators and should provide the following features.

1. Variable power, normally achieved via a dial setting, from zero to a maximum output of about 150 W cm^{-2} through a probe tip 2.5 cm in diameter. This will normally approximate to a quoted rating for the instrument of 500 to 600 Watts.

2. A pulse facility to allow intermittent sonication.

3. A range of interchangeable horns of the types described in Section 3.1 above. Replaceable tips (studs) for the 2.5 cm horn.

4. Horn extenders for the 2.5 cm system with screw thread fittings at both ends. Some extenders have only one end with a screw thread for extending the standard horn, but no screw fitting at the other end for further extension or for a replaceable tip.

5. Two useful attachments which extend the use of the equipment are a flow cell, to enable the probe system to be adapted for continuous processing (see Chapter 4), and a cup-horn (Section 2.7).

These probe systems are the 'work-horses' of sonochemistry, requiring only to be tuned to optimum performance in the system under study before being used to sonicate the system for time periods and at powers which are at the operator's discrection.

Additional desirable features

It has always been possible to have equipment built to particular specifications with extra options. Some of the extra options which are recommended in addition to those above are:

1. The option to use different frequencies. This will mean using a different transducer and horn for each required frequency and a different generator. For convenience the frequency generators can all be housed within one box with outlets for each probe system.

2. Automatic frequency tuning is useful, since the precise frequency match for the probe and the reacting system may not be exactly the nominal value for the transducer employed. A read-out of the optimised frequency is then important.

3. Some form of feedback to the generator from the transducer element, which can be used to maintain irradiation power throughout changes that occur during a reaction. A wattmeter reading of power supply to the transducer or vibrational amplitude of the horn is then an advantage.

3.3 Laboratory-scale sonochemical reactors involving probe systems

The simplest method of introducing power ultrasound to a reacting system via a probe system would be to dip the tip of the horn into the chemical reaction contained in a standard round bottomed flask or even a beaker. The temption then would be to turn the power to a maximum and wait to see what type of sonochemistry might develop. Unfortunately, this obvious approach is seldom ideal. Several precautions should be taken when designing a reactor involving a probe as a result of the following factors.

1. Volatile components may well be forced out of an open reaction by the 'degassing' effect of power ultrasound.

2. Probes can generate a great deal of heat and so cooling may be required.

3. When high-energy ultrasound is introduced into a liquid the power input may reach a stage where so much cavitation is produced that it blocks further power input (see Section 1.4). Thus the maximum setting of the generator power control does not necessarily produce the greatest sonochemical effect.

4. Mechanical mixing may be required at low-power settings. At higher powers the vibrating horn tip will produce sufficient streaming of the liquid away from the tip to produce effective bulk mixing.

5. Chemical reactions may well require vapour-tight apparatus, the slow addition of reagents, or inert atmospheres. Any sonochemical reaction of this type must be performed in a specialised vessel which accommodates such requirements.

Fitting a probe system into standard glassware

Commercial 20 kHz probe systems are supplied with standard 13 cm horns. Unfortunately, for many laboratory scale reactions, this is too short to enable the vibrating tip to reach the liquid medium. This problem is easily solved with the use of a half-wave extender (cylindrical horn).

Another problem is to find a way of sealing the horn into a reaction vessel. Here the difficulty is that at most points along its length the horn is undergoing ultrasonic vibrations. Fortunately there is a null point with a screw thread on most commercial sonicators on the booster horn (see Fig. 3.1). Anything screwed onto the horn at this point will not be subject to vibration, and so it is here that a PTFE sleeve adaptor can be attached which can be made to push-fit into a B34 cone of ground-glass apparatus (Fig. 3.3).

The design of reaction vessels

Once the horn has been sealed into the reaction vessel, some of the options then available to make it more suitable for ultrasonic use are given below.

The use of glass beads

A probe system will cause mixing in a reaction as a result of the liquid motion produced by the sonic waves generated at the horn tip. The mixing can be enhanced by putting glass beads in the bottom of the reaction vessel. The beads will be 'bounced' into motion by the sonic waves, and thereby assist bulk mixing.

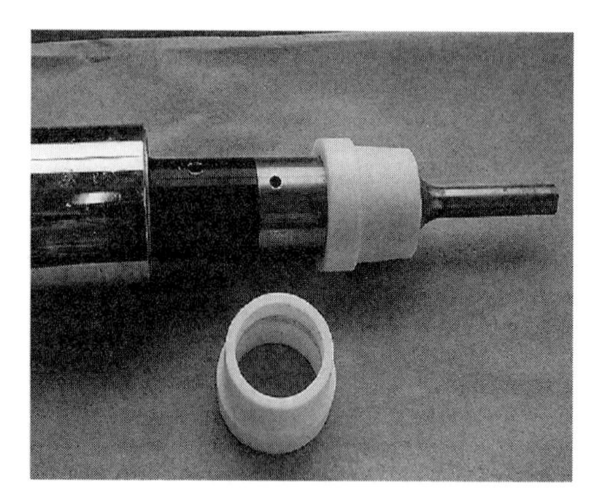

Figure 3.3 Probe fitted with B34 PTFE sleeve

Dimple cell

Another method of assisting sonic mixing is to adapt the reaction vessel by providing a bump on its base (Fig. 3.4). The probe should be positioned vertically above and no closer than 1 cm from the indent, which serves to disperse the sonic waves as they impinge upon and are reflected from the base.

Rosett cell

The Rosett cell was originally designed to be operated 'open-ended' for biological cell disruption. It allows the irradiated reaction mixture to be sonically propelled from the end of the probe around the loops at the base of the vessel, thus providing efficient mixing and also cooling (when the vessel is immersed in a thermostatted bath). When equipped with a flanged lid and a PTFE seal this is a good general purpose sonochemical reactor (Fig. 3.5).

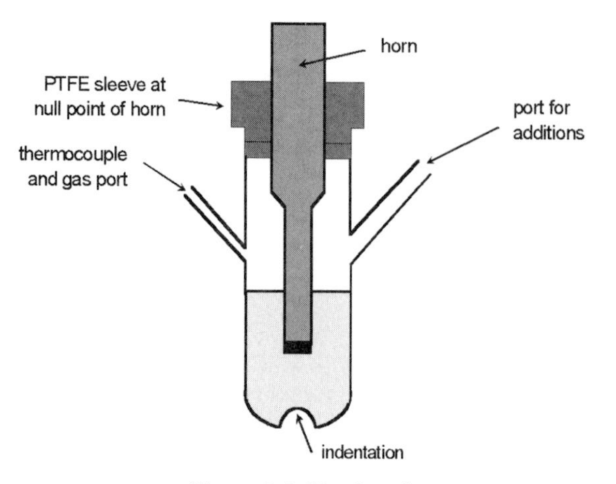

Figure 3.4 Dimple cell

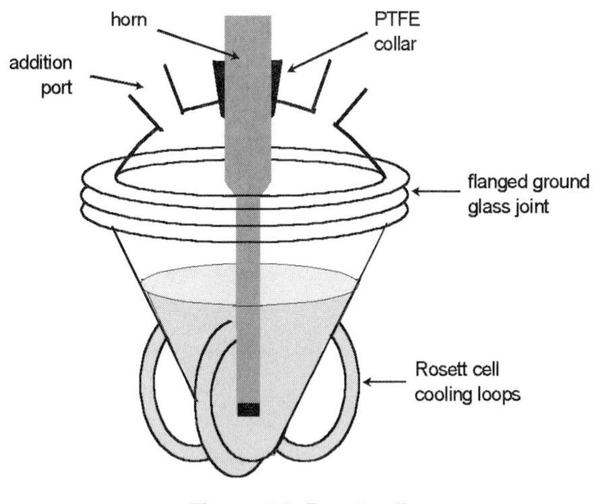

Figure 3.5 Rosett cell

Pressure cell

In situations where elevated pressures need to be used, a reactor has been described by Suslick. Here the screw thread on the horn is used to attach a PTFE collar which locks onto the outside of the glass joint of the sonochemical apparatus (Fig. 3.6). This provides a much more efficient pressure seal than the normal push fit of the PTFE sleeve into a ground glass joint.

A more refined solution has been developed by Undatim, a Belgian company. This cell provides an adjustable anvil to give the optimum reflection and therefore coupling of the ultrasonic power to the liquid system (Fig. 3.7).

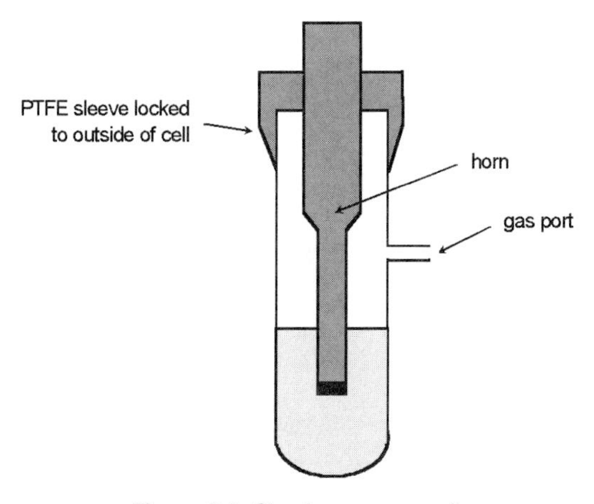

Figure 3.6 Simple pressure cell

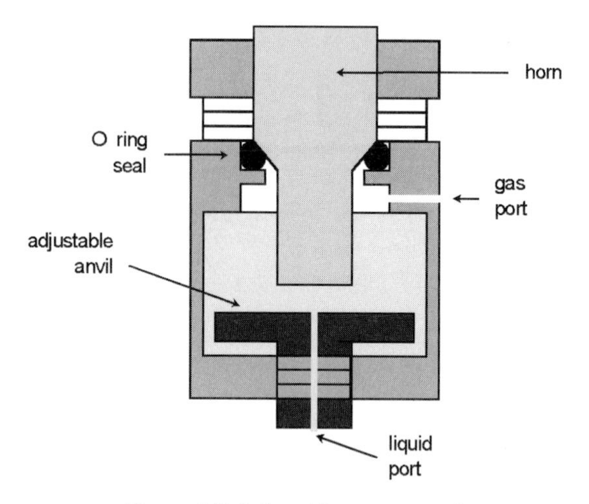

Figure 3.7 Adjustable pressure cell

Cooled cell with reagent holder

A piece of apparatus developed in the laboratories of J-L. Luche is shown in Fig. 3.8. External cooling is provided by a water jacket, and a small glass 'tray' is used to hold a metal reagent close to the horn tip, i.e. permanently in the maximum cavitation zone. This cell was originally used for the generation of dialkyl zincs.

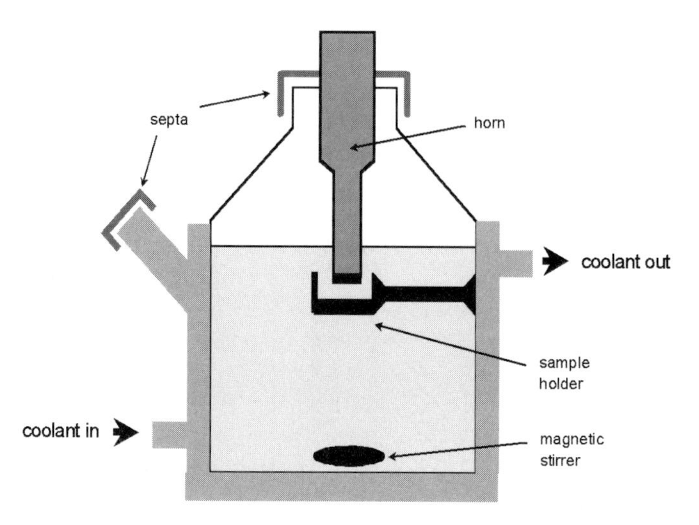

Figure 3.8 Cooled cell with reagent holder

3.4 The effect of external parameters on sonochemistry

There are a number of physical parameters which can be changed in order to affect the action of power ultrasound. This section gives the results of an investigation into how ultrasonic energy input to a chemical reaction is affected by the following important reaction parameters: ultrasonic power used, the presence of bubbled gas, temperature, solvent composition, and reaction volume. The work which follows in this section was performed using the Undatim Sonoreactor probe system which is equipped with an automatic transducer resonance frequency search device. This enables the power input to a system to be maintained accurately throughout a reaction, thereby offering considerable advantages compared with basic commercial probe systems in terms of both the monitoring and control of irradiation parameters.

The type of reaction system

Of the many different classes of reaction which can be enhanced by ultrasonic irradiation the most common are heterogeneous, including both liquid/liquid and liquid/solid. The chosen system for study was the particle size reduction of copper bronze in liquid media under a variety of different conditions [1]. Although this is not a 'chemical' reaction as such, the reduction of particle size, i.e. increase in available surface area of a reactant or catalyst, is a major contributor to sonochemical enhancement of solid/liquid reactions. Moreover the size reduction is a readily quantifiable factor and thus is ideal as a guide to estimating sonochemical efficiency for the range of conditions chosen. It is important to note that the results obtained in this study are based upon a solid/liquid heterogeneous model reaction, and so may not be directly applicable to heterogeneous liquid/liquid or homogeneous reactions for which different models may be necessary.

The effect of bubbled (entrained) gas

If the main cause of sonochemical effects is cavitational collapse, then it is clear that sonochemical efficiency must be linked directly to the generation of cavitation bubbles in the chemical reaction. For reproducible results many workers have used a bubbled gas to generate large numbers of nucleation sites for cavitation and provide bubbles of uniform energy of collapse. Most effective for such processes are the monoatomic gases such as argon, with diatomics such as nitrogen next, and polyatomics a very poor third. This is directly related to the ratio of specific heats for these gases—the highest ratios giving the greatest cavitational effects.

The particle size reduction of copper bronze, induced by sonication, illustrates the effect of entrained gas on sonochemistry (Table 3.1). The results clearly show that, if all other external parameters are constant, there is a distinct advantage in the continuous passage of argon through the system.

The only disadvantage of such methodology in general sonochemistry is that, at reaction temperatures above ambient and in an open system, there could well be losses of volatile material through evaporation. It has been

Table 3.1 The effect of entrained gas on the particle size reduction of copper bronze in water at 30 °C (20 kHz)

Entrained gas	Particle size after 1 hour (μm)
No gas	23.8
Nitrogen	22.9
Argon	19.1

Original particle size 60 μm, intensity 65 W cm^{-2}, concentration 2 g in 100 cm^3 water.

demonstrated that efficient saturation of a gas with liquid vapour is achieved by insonation of that liquid with the gas passing through [2]. For this reason in the studies reported below the sonication has been performed in the absence of entrained gases. Except where stated, all reactions were performed using a 'dimple cell'.

The effect of changes in the reaction medium (the liquid load)

As the temperature of a reaction is increased, the vapour pressure of any liquid involved in the reaction is also increased. The collapse energy of a cavitation bubble, formed during sonication of the medium, will be directly influenced by the vapour which enters the bubble during its formation. The more vapour which enters the bubble the more cushioned is the collapse, so that any increase in the temperature (i.e. vapour pressure) of a system will reduce the energy of cavitational bubble collapse. Thus as the system temperature is raised, the ultrasonic power input will need to be increased if a relatively constant sonochemical effect is to be maintained. Alternatively if the ultrasonic power setting on the generator is constant, then the lower the temperature of operation the greater will be the sonochemical effect. This is demonstrated in Table 3.2, where the probe system operating at 20 kHz was set at a constant power rating of 7 during the insonation of water at four different temperatures. The effect of reducing the bulk water temperature from 90 °C to 0 °C, in steps of 30 °C, on the reduction in particle size of copper bronze powder, shows clearly the improvement achieved as the temperature is lowered. The intensity column shows how the measured power entering the system changes with temperature at the same instrument setting.

Table 3.2 The effect of temperature on the particle size reduction of copper bronze in water at 30 °C (20 kHz) at same instrument power output setting

Temperature	Intensity (W cm^{-2})	Particle size after 1 hour (μm)
90	23	31.9
60	39	29.3
30	62	23.8
0	79	20.6

Original particle size 60 μm, concentration 2 g in 100 cm^3 water.

The effect of reaction temperature and solvent composition on irradiation frequency and power

If cavitation is the major cause of sonochemical effects, then factors such as the vapour pressure, viscosity, and surface tension of the solvent will have a bearing on sonochemical reactions. Any change in the reaction temperature results in a change in these parameters with, as a consequence, a change in the optimum resonance frequency for the system. For example, employing the frequency search mode of the Undatim instrument during the irradiation of 100 cm^3 of

Table 3.3 Optimum resonance frequencies at different temperatures using probes of different nominal transducer frequencies

Nominal frequency (kHz)	Temperature (°C)	Optimum resonance frequency (kHz)
20.000	30	20.668
–	60	20.627
–	90	20.578
40.000	30	39.262
–	60	39.128
–	90	39.024
60.000	25	59.652
–	50	59.486

Results obtained in 100 cm^3 water in a glass beaker.

water at different temperatures and employing three different transducers, a small but significant change in frequency was observed (Table 3.3). The precise change depends upon the geometry of the vessel and the position of the probe tip.

All chemical reactions, by definition, proceed to a product which differs from the starting material, i.e. the properties of the reaction mixture change with time, therefore, there will be a change in the acoustic properties of the medium during the course of a reaction. As a model for this change 200 cm^3 of water and dimethylformamide (DMF) at 20 kHz were separately sonicated and the power input to each system was monitored. At instrument setting 7 it proved possible to transmit 122 watts of power to the aqueous system, but in DMF the power dissipated at the same setting was only 101 watts. Thus, in order to maintain optimum performance for any sonication system where the reaction temperature is not constant, it may be necessary, with conventional instrumentation, to adjust the tuning of the transducer during the course of the reaction (see Section 3.1).

The effect of changes in the sonication frequency

One of the problems associated with operating at different insonation frequencies is that the physical dimensions of the transducers and horns place a real limit upon the power (i.e. amplitude of vibration) which they can deliver. A comparison of the amount of energy (calorimetric) dissipated in 100 cm^3 of water at 30 °C by three different horn systems is shown in Table 3.4. In the same table the physical consequences of this can also be illustrated experimentally by observing the rate of particle size reduction for copper bronze at 30 °C.

It should not be concluded from these results that the optimum sonochemical effect can only be obtained at high powers, i.e. at low (20 kHz) frequency. From these results it is certainly tempting to assume that the greatest effects are achieved at lower frequencies. What must be borne in mind, however, is that the actual ultrasonic power entering the system also depends

Table 3.4 The effect of change in irradiation frequency on optimum power input into a system

Nominal frequency (kHz)	Tip area (cm^2)	Calorimetric power (W)	Intensity ($W\ cm^{-2}$)	Particle size (μm)
20	1.40	96	68.6	23.8
40	1.00	26	26.0	37.7
60	0.75	11	14.7	43.2

Original particle size 60 μm, concentration 2 g in 100 cm^3 water, sonication time 1 hour at 30 °C.

upon the frequency—high frequencies are delivered at lower powers. Thus in order to properly equate the effects of ultrasonic irradiation at different frequencies the time of exposure must be considered. A report on the de-agglomeration and dispersion of SiC has effectively made this point [3]. In order to de-agglomerate a material with ultrasonic irradiation tow main parameters must be considered:

1. A threshold power level will be necessary to obtain any size reduction and may depend on the frequency.

2. Whatever frequency is used, it is the total sum of ultrasonic energy delivered to the system which is important.

Thus one might expect that the degree of de-agglomeration will be dependent on the total energy (time × power), and results should be more meaningful if the data are compared using this measure of the total energy. The effect of producing the same total energy from a system using 20, 40, and 60 kHz probes can be seen in Table 3.5.

The results clearly indicate that even if the same total energy is used the 20 kHz probe system does appear to give a better size reduction than the 40 or 60. Significantly, however, both the 40 and 60 kHz systems give very similar results. The same experimental comparison was also made using two different powders, aluminium oxide and nickel, both suspended in water. Again, the same trend was seen. The 20 kHz probe produced a greater particle size reduc-

Table 3.5 Particle size reduction achieved at different frequencies at same total energy input (Power × time) of 480 W min

Frequency (kHz)	Calorimetric power (W)	Time (min)	[Particle size (μm)]		
			[Cu]	[Ni]	[Al_2O_3]
20	96	5.0	36.44	51.00	64.50
40	26	18.5	43.46	56.64	70.96
60	11	43.6	44.17	57.96	74.01

Temperature 30 °C, 100 cm^3 water in a dimple cell, original particle size Cu 60 μm, Ni 70 μm, Al_2O_3 85 μm.

tion than the 40 or 60 kHz probe at the same energy delivered. The effect of insonation frequency in sonochemistry is, however, a relatively new area of study, and indications are that in homogeneous reactions and those involving emulsification there may well be a significant frequency effect.

3.5 The advantages and disadvantages of using an ultrasonic probe system for sonochemistry

In the chemical laboratory the commercially available probe system is the best method of introducing high-power ultrasound into a chemical reaction. Its advantages and disadvantages are summarised below.

The advantages of using an ultrasonic probe system for sonochemistry

1. The ultrasonic power delivered by a horn is directly related to the magnitude of vibration of the tip. This can be readily controlled by the power input to the transducer, and so the precise power of the system can be regulated. Maximum powers of several hundred W cm^{-2} can be easily achieved (depending on the size of the unit).

2. Ultrasonic streaming from the tip of the probe operated at moderate power is often sufficient to provide bulk mixing.

3. Most modern units have a pulse facility allowing the operator to sonicate reactions repeatedly for fractions of a second. This gives adequate time for 'cooling' between sonic pulses.

4. With such systems the probe system can be tuned to give optimum performance. This is important in terms of reproducibility of results. Modern equipment is normally fitted with an Automatic Frequency Regulator.

The disadvantages of using an ultrasonic probe system for sonochemistry

1. As with all systems which operate using piezoelectric transducers, the optimum performance is obtained at a fixed frequency. For most commercial probe systems this is 20 kHz and, although it is possible to drive them at their overtones (i.e. 40 or 80 kHz), the power dissipation at overtones of the fundamental frequency of the system is very much reduced. To operate successfully at different frequencies it is best to purchase individual amplifier/horn systems tailored to individual requirements.

2. As with baths there is a problem over accurate temperature control unless precautions are taken. The use of specially designed reaction vessels, e.g. a Rosett cell, does alleviate much of this difficulty. The temperature within the reaction vessel should be continuously monitored.

3. Due to the high intensity of irradiation in the reaction close to the tip, it is possible that radical species may be produced which could interfere with the normal course of reaction.

4. The cavitation which is the source of chemical activation is also the source of a common problem with probe systems—tip erosion—which occurs despite the fact that most probes are fabricated of titanium alloy. There are two unwanted side effects associated with erosion: (i) metal particles eroded from the tip will contaminate the reaction mixture and (ii) the physical shortening of the horn causes a loss of efficiency (eventually it will become too short to be tuned). The latter problem is avoided by the use of screw-on tips to the probe in the form of studs, this eliminates the need for a costly replacement of the whole horn.

5. Special seals will be required if the horn is to be used in reactions involving reflux, inert atmospheres, or pressures above (or below) ambient.

6. Laboratory probe systems are generally only suitable for small batch reactions, although multiple probes will cope with larger volumes.

References

1. T. J. Mason, J. P. Lorimer, and D. M. Bates, *Ultrasonics* (1992) *30*, 40.
2. T. J. Mason and P. Sephton, *Aldrichimica Acta* (1989) *22*, 2.
3. M. Aoki, T. A. Ring, and J. S. Haggerty, *Advanced Ceramic Materials* (1978) *2*, 209.

4 Flow systems and scale-up in sonochemistry

It has been recognised for many years that power ultrasound has great potential for use in a wide variety of processes in chemical and allied industries (Section 1.2). The interest of chemists in power ultrasound has also been established, whether it be to improve yields, speed up chemical or physicochemical processes, or, as seems possible in some cases, completely switch reaction pathways. Yet, in spite of the great amount of information available on sonochemistry, and in spite of the very promising effects described on a laboratory scale, the literature does not seem to provide a great many examples of industrial applications. Over the last few years a number of groups have become interested in solving the problems of scale-up, and the purpose of this chapter is to give an overview of the available systems that can be used to carry out sonochemical transformations on a scale rather larger than that used in the research laboratory [1,2].

4.1 The challenge of scale up

There are essentially two types of large-scale chemical plant: batch and flow type. Sometimes the flow system will form a part of batch processing as a loop attached to the main vat. In the preceding chapters we have concentrated on laboratory-scale equipment. The results from some of these small-scale experiments can be adapted for large-scale work, but in order to do so a precise knowledge of the type of sonochemistry involved in the process is essential.

The first decision to be made when considering how a sonochemical reaction should be scaled up is whether that reaction is in fact truly sonochemical or simply the result of an ultrasonically induced mechanical effect. In a solid/liquid heterogeneous reaction, for example, power ultrasound may only serve as an efficient mixing system (for particle fragmentation, deagglomeration, and/or dispersion). In this situation ultrasonic pre-treatment of a slurry may be all that is required before the reaction is allowed to take place conventionally. However, if ultrasound has a real effect on the chemistry of the system, then arrangements must be made to provide sonication during the reaction itself. To scale-up a true sonochemical process, there are three main questions to be answered:

1. What will be the most appropriate sonolysis system for the reaction mixture to be processed?

2. What are the best reaction conditions?

3. What will be the energy implications of using ultrasound rather than traditional methodology?

These questions are not easily answered because of the number of factors which must be taken into account when judgement is made. Thus, to attain optimum conditions for sonication the variables which can influence cavitation must be carefully studied. These variables depend upon:

1. The reaction medium: in terms of its viscosity, vapour pressure, the nature and concentration of any dissolved gas, and the presence of any solid particles before, during, or after reaction.

2. The reaction conditions: including the temperatures and pressures involved—these may well be varied during a conventional process.

3. The type of sonic system used: this must involve consideration of both the power and the best frequency to be used, together with the size and geometry of the chemical reactor to be employed.

4.2 Sonochemical efficiency

Table 4.1 Methods of introducing ultrasound into a reaction

Methods of introducing ultrasound into a reaction
Dipping a vessel containing the reaction mixture into a tank containing a sonicated liquid (most generally water)
Using a reaction vessel whose walls are subjected to ultrasonic vibrations
Immersing a source of power ultrasound into the reaction medium itself

There are only a limited number of ways in which power ultrasound can be introduced into a chemical reaction (Table 4.1). Whichever method is employed it is certain that there will be a loss in efficiency of the process between the mains electrical power feeding the generator and the acoustic power consumed to perform the chemical process in the reaction. The first place where any loss in efficiency can arise is in the design of the generator where mains power is converted into a signal to the transducer. At this stage there is an inevitable loss of heat. The motion of the transducer will itself involve further losses of energy in the form of heat and sound. Further inefficiency arises from the 'coupling' of the transducer to its load, i.e. the mass which it is sonicating. Lastly there is the problem of whether it is in fact possible to achieve sonication throughout the whole reaction mixture, or whether it would be easier to localise the power ultrasound into a particular region of the reacting system.

Figure 4.1 can be considered to represent a sonochemical process from the energetics viewpoint. The ultrasonic system transforms electrical power into vibrational energy, i.e. mechanical energy. This mechanical energy is then transmitted into the sonicated reaction medium. Part of it is lost to produce heat, and another part produces cavitation, but not all of the cavitational energy produces chemical and physical effects. Some energy is reflected and some is consumed in sound re-emission (harmonics and sub-harmonics). From this, it can be appreciated that the determination of an energy balance is not easy. Nevertheless one can define the term 'sonochemical yield' (SY), as:

$$\text{SY} = \text{'measured effect'}/\text{input power in watts}$$

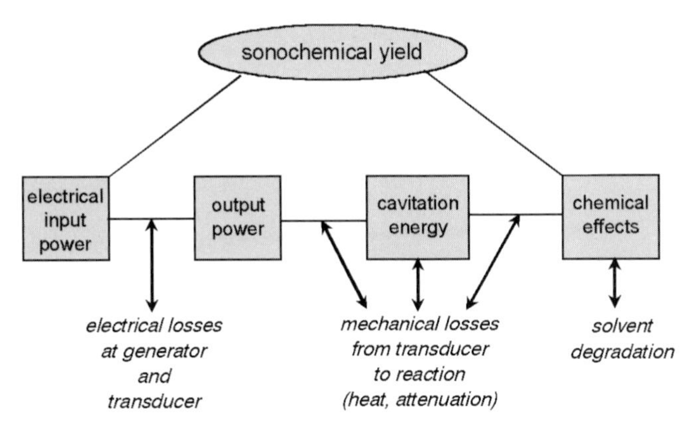

Figure 4.1 Possible energy loss during a sonochemical reaction

This is similar to the definition of 'photochemical yield'. The measured effect can be, for example, the number of moles of product generated per second and the input power can be monitored and measured with a wattmeter. To estimate the output power, i.e. the actual power entering the system, one has to know the conversion yield of the ultrasonic system used. This is not always obvious since part of the energy is reflected back to the transducer, especially when the sonicated mixture is viscous or contains a large amount of suspended solid. Several attempts have been made to measure the overall output power with calorimetry and several different chemical dosimeters (see Chapter 1).

Another problem is the determination of the way in which the 'sonic energy' is distributed within a particular reactor configuration using a given sonic system. One approach developed for the determination of this distribution involves thermistor probes [3]. Pugin has shown that the distribution of ultrasonic energy is not homogeneous (Section 2.3) [4]. In some cases standing waves may be formed, and there is a question as to whether a larger sonochemical effect is obtained with standing waves or in their absence [5]. The energy distribution might well have some effect on the course of a chemical reaction. Consider the case of a solid–liquid phase transfer reaction (Scheme 4.1) [2]. In this Michael addition of the anion of pentan-2,4-dione (2) to chalcone (1) the adduct (3) is formed and this can react further to give the cycloadduct (4). The overall rate of this reaction is increased by sonication, but the chemical selectivity depends on the particular source of ultrasonic power. Thus, adduct (3) is the only product formed in the absence of ultrasound or using a cup-horn, whereas the cyclised adduct (4) is produced only under the influence of sonication with a probe (Table 4.2). The reason is probably that at the surface of the horn (diameter 12 mm) the 'power density' (expressed in W cm^{-2}) is higher than at the surface of the cup-horn (diameter 45 mm). This would result in far more severe conditions in the small volume directly under the horn, which might then promote the cyclisation side-reaction.

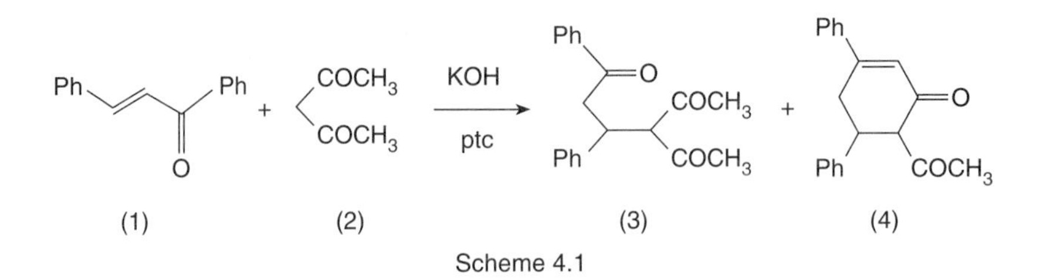

Scheme 4.1

Table 4.2 Effect of the source of ultrasonic energy on percent production distribution in a condensation reaction

No sonication		Cup-horn		Probe system	
(3)	(4)	(3)	(4)	(3)	(4)
52	0	69	0	72	12

This difference in product distribution points to one obvious problem when scaling-up sonochemical processes. In an ideal situation the correct sonication conditions would be established on a laboratory scale and the input power required would be optimised for the best sonochemical yield. The energy input could then be characterised as the energy distribution in the reactor, either as the energy emitted at the surface of the ultrasonic device (expressed as W cm^{-2}) or the energy dissipated in the bulk of the sonicated medium (expressed as W cm^{-3}). In the scaled-up version of the reactor, the shape and type of device may not be directly comparable to the laboratory reactor, and so product differences may arise.

This is less likely to be a problem when dealing with smaller sonicated volumes such as those used with flow cells, where the geometries and ultrasonic sources are similar on both scales.

4.3 Large-scale ultrasonic systems

Solutions to the problem of the scale-up of sonochemical reactions do exist but they are not so simple as the use of bigger versions of laboratory equipment. In a production situation the volumes treated will be very much larger than those considered in the laboratory, and the type of process will govern the choice of reactor design. It could well be that some processes would be more suited to low-intensity sonication (e.g. using a bath-type reactor or a liquid whistle), whereas others may need higher intensity irradiation of the type obtained using a probe-type system.

One of the major considerations which have held back progress on the development of large-scale sonochemical equipment has been that industrial sonochemistry might require 'tailor-made' reactors to achieve optimum sonochemical yields for specific reactions. Such reactors would clearly involve considerable capital investment. Fortunately, however, a wide range of equipment is already available for processing and cleaning which can be readily adapted for sonochemistry (Table 4.3).

Table 4.3 Commercially available sonication systems

Cleaning bath (including submersible transducers)
Probe systems (including flow cell, cup-horn)
Shaking tray
Liquid whistle
Flow reactor (various cross-sections)
Radial vibrating tube

Batch reactors

Based on the ultrasonic cleaning bath

In cases where only low-intensity ultrasonic irradiation is needed, batch treatment could well be as simple as immersing the reaction vessel in a large-scale ultrasonic cleaning bath. This method has been adopted for a medium-scale Simmons–Smith cyclopropanation reaction using sonochemically activated zinc [6]. Most conventional methods for carrying out this reaction rely upon activation of the zinc by using zinc–silver or zinc–copper couples and/or the use of iodine or lithium. A sudden exotherm and excessive frothing are common practical problems. In the sonochemical procedure no special activation of the zinc was required, indeed equally good—and reproducible—yields were obtained using zinc in the form of dust, foil, mossy, or even rods. This methodology was successfully scaled up to achieve the cyclopropanation of methyl oleate in 0.5 kg quantities (Scheme 4.2).

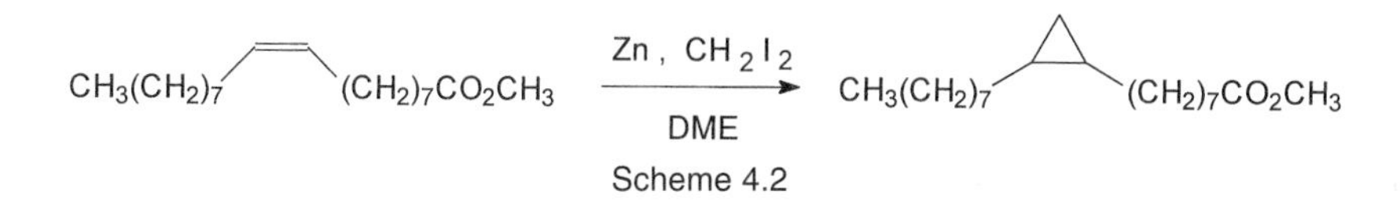

Scheme 4.2

The reactor (shown in Fig. 4.2) is a 22 litre flask immersed in a 36 × 18 × 18 inch, 50 gallon heated ultrasonic bath (3 kW, 80 kHz). The zinc metal was cast in the form of two cones (each 900 g) which were suspended with copper wire thus allowing them to be moved into or out of the reaction mixture. The

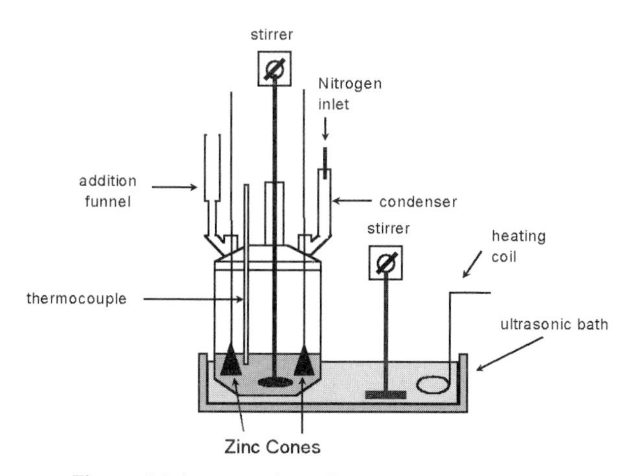

Figure 4.2 Large scale cyclopropanation apparatus

method had several advantages over the normal method of cyclopropanation as a result of changing from zinc powder to the metal:

(i) there was a reduction in foaming (normally associated with ethene and cyclopropane formation);

(ii) the exotherm was more evenly distributed (only a small clean area of metal is available throughout the reaction);

(iii) the reaction could be controlled by removing the lump of metal from the reaction; and

(iv) the residual metal could be removed from the reaction as a lump rather than by conventional filtration.

If the cleaning bath itself is to be used as the reaction vessel, the bath tank itself would need to be constructed of a material which was inert towards the chemicals involved. An appropriate grade of stainless steel might prove adequate or plastic tanks could be used. In the latter case, however, the transfer of ultrasonic energy from the tank walls to the reaction mixture would be attenuated by the plastic, and so the transducers would need to be bonded to a stainless steel or titanium plate and this assembly then bolted to the inside wall of the tank. In addition the tank could well require a sealed lid with addition ports, and almost certainly stirring would be required.

Reaction vessel with external transducers

A more logical use of this concept would appear to be to simply convert an existing reaction vessel by the addition of external transducers (Fig. 4.3). In this configuration the walls of the vessel provide the ultrasonic energy in the same way as the walls of a cleaning bath. This system would have the advantage that facilitates for stirring and for the use of elevated temperatures and pressures would already be available.

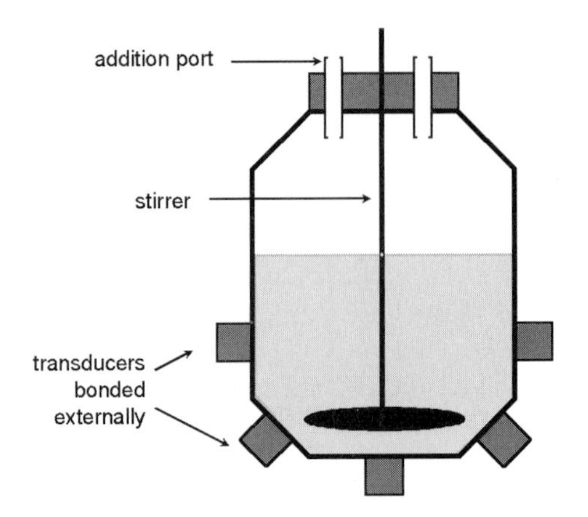

Figure 4.3 'Cleaning bath' reactor

There are, however, some major disadvantages to this type of adaptation.

1. The transfer efficiency of the acoustic energy entering the reaction will be quite small and the presence of a stirrer or solids will cause attenuation of the sound energy, and this may lead to a central zone which is not under sonication. This will place a limit on the maximum size of the vessel. In this configuration the energy distribution will also be hard to determine.

2. If the surfaces to which the transducers are to be attached are curved then very careful design will be required to attain the best energy transfer.

3. For heated reactions the transducers will need to be cooled and/or removed from direct contact with the vessel walls. The use of cylindrical extension horn (Section 3.1) will enable the transducers to be positioned away from the vessel walls at some multiple of the half-wavelength of the sound in the horn material.

Bath-type reactors can also be used in a flow system. In this case the reacting liquids could be continuously fed into an ultrasonic tank with outflow to the next process. Such treatment could be intensified by recycling or by connecting a number of such sonicated tanks in line.

The submersible transducer

A useful variant to the sonicated tank is the submersible transducer assembly described above (Fig. 2.6). In fact this offers somewhat greater flexibility in use, since it comes as a sealed unit capable of withstanding organic solvents—this dates from the days when halocarbon-type cleaning fluids were in common usage. Such units can be designed to fit into any existing reaction vessel. As with the cleaning bath, ultrasonic reactors involving submersible transducers would require some form of additional (mechanical) stirring.

Fluidised bed reactor

A sonochemical reactor has been designed which utilises an externally cooled conical shape vessel with a probe at the base (Fig. 4.4) [7]. The system is of the cup-horn type except that the cup is in the shape of a cone constructed with an internal angle which allows optimum radiation of the acoustic field giving plane waves in the reaction chamber. All fluid reactants enter the reaction chamber via a concentric inlet around the probe, and the fluid motion keeps reacting solids suspended in the active sonication zone. In essence the system operates rather like a fluidised bed allowing liquids to mix intimately with suspended solids thereby giving very efficient sonochemical conditions. One obvious application area for this system is Grignard syntheses where the magnesium metal is the suspended material.

Batch reactors incorporating a probe system

The ultrasonic probe system, with the various types of horn design available to amplify vibrational amplitude (see Chapter 3), is one of the most effective

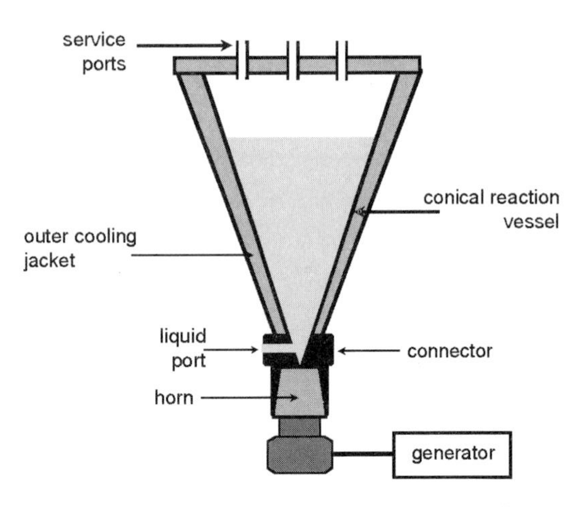

Figure 4.4 Conical reactor with fluidised bed effect

methods of introducing high-power ultrasound into a liquid. Despite this there are some severe restrictions in usage for large-scale batch treatment:

1. The high energy density produced from a single probe tip is unlikely to be capable of delivering sufficient energy density to affect the whole of a large reacting volume, despite the circulation induced by streaming. Unlike the bath system described above, a probe reactor tends to have its 'dead zones' on the periphery of the vessel (Fig. 4.5).

2. In some reactions an optimum power level can be reached (which depends on the system) beyond which increased power input actually reduces the yield (see Section 1.4).

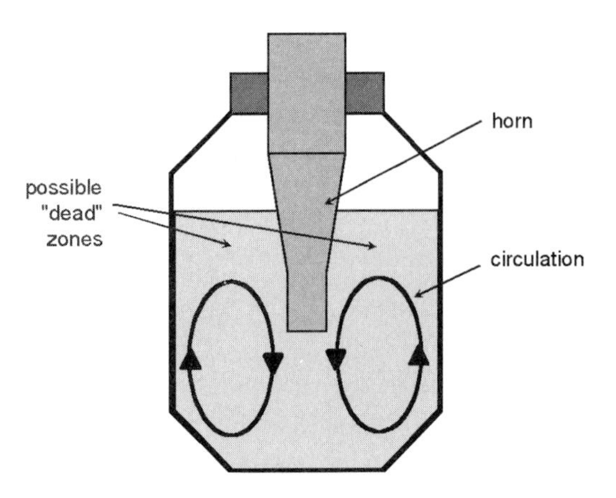

Figure 4.5 Probe batch reactor

3. At high power, erosion of metal from the probe tip could contaminate the reaction and may also necessitate frequent probe or tip replacement. High cavitation energies may also cause some unwanted radical formation and/or solvent decomposition.

Flow systems

For larger scale operations where high intensity ultrasound is required, the sonic horn, or indeed any other source of high power ultrasound, is best utilised as part of a flow loop outside of the main reactor (Fig. 4.6). This permits the processing of large volumes and provides significant advantages compared with the use of probes in a batch configuration. In a flow system, high intensity sonication can be provided at controllable power by the adjustment of either input power to the transducer system or the flow rate through the cell. In addition temperature control is provided through the circulating reaction mixture or using a pulse option. There are also some disadvantages to the use of a flow system, the most common of which is that pumping is required for circulation through the flow loop and so it is not suitable for either very viscous reaction media or heavily articulate systems.

In the schematic flow system shown the ultrasound generator can be one of several different configurations (Table 4.4). These possibilities are explored below.

Table 4.4 Ultrasonic processors for use in flow loops

Flow cell with probe
Resonating tube reactor of different cross-sectional geometry
Cylindrical resonating bar inserts
Liquid whistle
Vibrating tray

Flow cell attached to a horn system

A simple option for the initial investigation of flow treatment is the flow cell which is generally available as an option for laboratory probe systems (Fig. 4.7).

There are certain disadvantages to using a probe in the flow system apart from the problems with medium viscosity and particle circulation. These have already been mentioned for probe systems themselves, i.e. high power

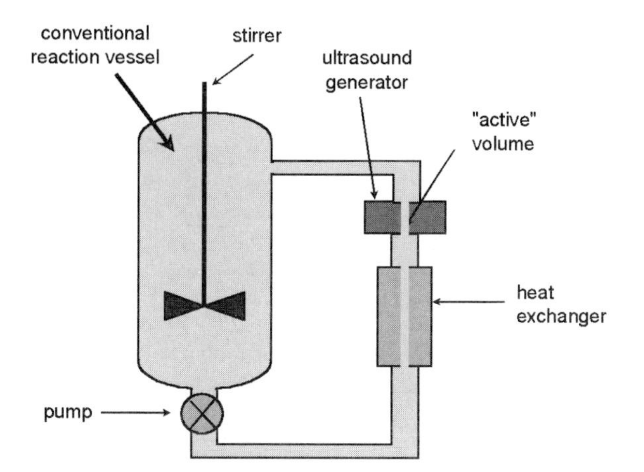

Figure 4.6 Schematic flow loop system

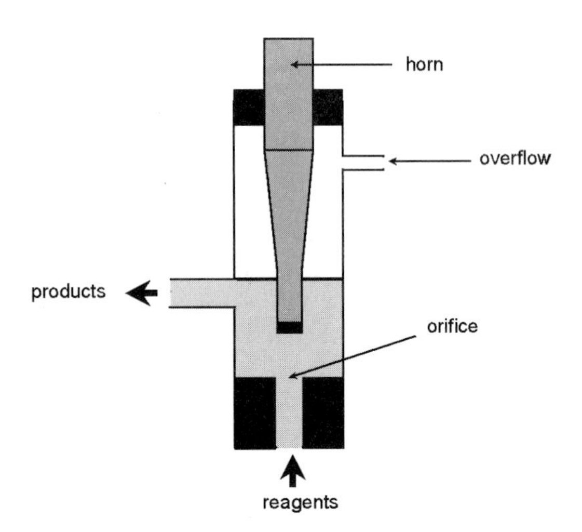

Figure 4.7 Flow cell attachment for probe system

treatment can produce tip erosion, some solvent decomposition, and the generation of radical species.

Resonating tube reactors

This type of system provides the greatest hope for general usage of ultrasound in the chemical industry. Essentially the liquid to be processed is passed through a pipe with ultrasonically vibrating walls. In this way the sound energy generated from transducers bonded to the outside of the tube is transferred directly into the flowing liquid. Two design engineering problems are associated with this type of sonicator (i) the correct mounting of the transducers on the outer tube and (ii) the length of the tube must be such that the ends are at nodal points in the sound wavelength in the unit. This will eliminate vibrational problems associated with the coupling of the unit to existing pipework. Generally the commercial tube reactors are constructed of stainless steel, for chemical processing, however, other materials such as titanium can be used.

Each of the following systems requires that ultrasonic energy is introduced into a flowing system by vibrations induced in the walls of the pipe carrying the reaction medium. With the correct length and fittings these flow reactors may be easily retrofitted to a chemical processing rig. The cross-sectional geometry for the pipe is generally one of the four shown (Fig. 4.8). Each of these has been used and each has its own good and bad points, but in general the bonding of transducers to a flat surface is easier than to a cylindrical tube.

(i) *Rectangular cross-section.* The transducers face each other in this geometry and so there is the possibility of severe erosion of opposite faces unless the design is optimised. If the pipe is not seamless, then high powered ultrasonic vibration is likely to find weaknesses in the construction, particularly at corner joints.

rectangular unfocused

pentagonal unfocused

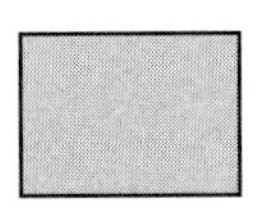

hexagonal focused

circular focused

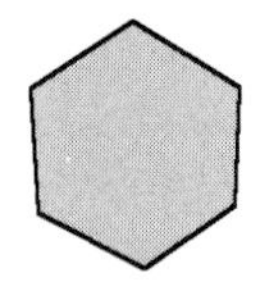

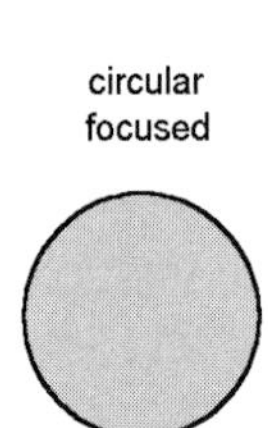

Figure 4.8 Cross-sectional geometries for flow reactors

One application for such a sonicated tube is in waste water treatment where large quantities of sewage sludge are produced which mainly consist of flocculent microorganisms. Anaerobic fermentation is the most commonly applied process for stabilization of sewage sludge, but this is a slow process with residence times in anaerobic digesters of about 20 days. A rectangular flow processor operating at 3.6 kW and 31 kHz at an intensity of 7.5 W cm^{-2} has been developed to pre-treat and disintegrate the sludge before digestion [8]. This reduced the residence time in the fermenter to 8 days thereby permitting a substantial increase in throughput.

An alternative type of rectangular reactor employs opposite emitting faces with different ultrasonic frequencies known as the Nearfield Acoustic Processor (NAP) (Fig. 4.9). This system can be visualised as two sonicated metal plates—each similar to the base of a cleaning bath—these enclose a flow system. The plates are driven at different frequencies—normally 16 and 20 kHz. In effect the plates can be regarded as the bases of two ultrasonic baths facing toward each other and separated by adjustable spacers. Under these conditions any liquid flowing between the plates is subject to an ultrasonic intensity greater than that expected from a simple addition of the single plate intensities. The ultrasound 'reverberates' and is magnified in its effect. An additional benefit is that with vibrating plates the size of table tops, the system can cope with a very large throughput of material.

(ii) *Pentagonal cross-section*. In this type of system the acoustic field is fairly even throughout the cross-section because the emitted wave does not impinge directly on the opposite face (Fig. 4.10). The dimensions of such a system are dictated by the space requirements for the transducers. Thus using 40 kHz transducers the cross-section has a minimum value of approximately 9 cm, and such a system with a volume of 1.5 litres has been used for processing [2]. Applications for this type of processor include pasteurisation, crystallisation, degassing, dispersion, and emulsification.

(iii) *Hexagonal cross-section*. A reactor of this type provides a 'focus' of energy in the centre of the pipe—unlike the pentagonal style. Such a focus

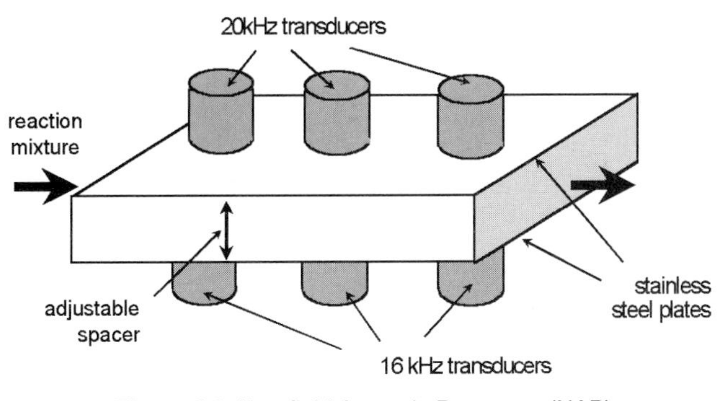

Figure 4.9 Nearfield Acoustic Processor (NAP)

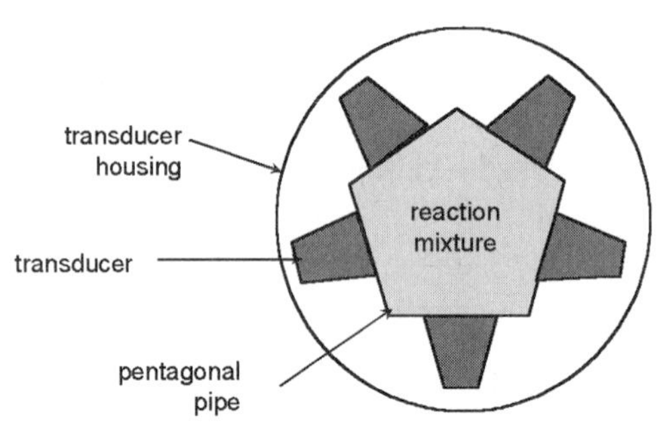

Figure 4.10 Pentagonal reactor

means that very high intensities can be developed away from the tube walls without the associated problems of rapid cavitational erosion on the inner surface of the walls. When the reactor is produced in the form of an ultrasonic bath, the reaction medium can be located in a vessel at the focus of the energy either for batch treatment or as the sonochemical reactor of a flow system (Fig. 4.11).

(iv) *Cylindrical cross-section.* To make a permanent bond for a high power transducer to a curved surface is not easy. Fortunately the geometry of the tubular configuration, just as with hexagonal, means that low power can be used at the surface because the ultrasonic energy is focused toward the middle of the tube. A resonating pipe of 6 inch diameter operating at 25 kHz has been developed for the degassing of liquids [9]. Another has been installed on offshore oil drilling rigs for the de-agglomeration of drilling mud. This process is required so that the jettisoned mud disperses with the movement of the sea and does not simply sink to the seabed under the plat-

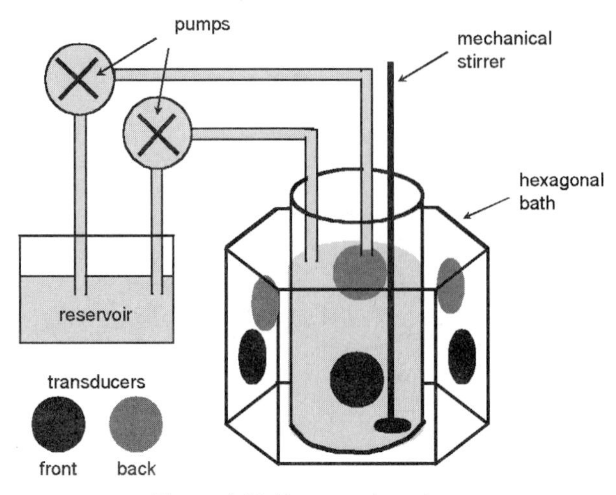

Figure 4.11 Hexagonal reactor

Figure 4.12 Large scale tube processor

form. Accumulations of mud around the platform legs present a problem to sea life and eventually rig maintenance. The tubular reactor meets strict standards of operation for use in hazardous areas, and consists of four modules each one rated at 2 kW and composed of a stainless steel tube of 1 metre in length and 12 cm diameter with an array of piezoelectric transducers bonded to the outside and enclosed in a casing (Fig. 4.12) [10].

An alternative approach to generating a vibrating pipe is to use a fluid to couple energy from a transducer to the outer wall of the reactor. This avoids any direct bonding to the surface. Such a reactor was developed from an ultrasonic wire cleaning system (Fig. 4.13). Transducers transfer energy to the cylindrical reactor tube via a coupling fluid. This same coupling fluid can be circulated and refrigerated to provide cooling during the sonochemical processing.

When a probe system is used as the source of energy and coupled through a fluid to a pipe, the advantage of controllable power is added to temperature control (Fig. 4.14). Like all systems based on probes, however, this unit will suffer from erosion of the horn tip and, in the long run, this might prove expensive in terms of 'down-time' for repair and replacement.

When very high ultrasonic energy is applied to a fluid through a pipe one unavoidable problem is the generation of heat. This can be overcome if the

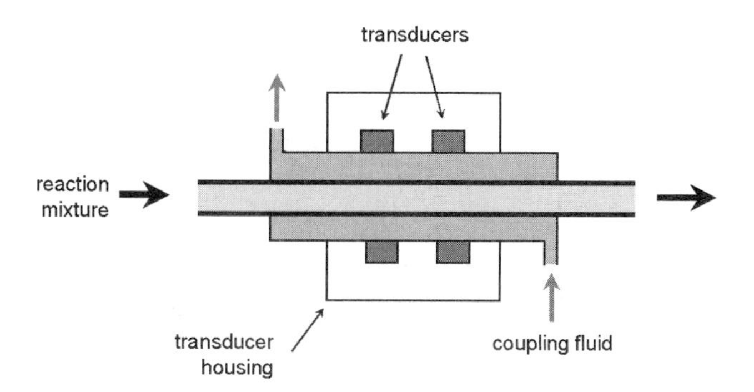

Figure 4.13 Tube reactor with liquid coupling

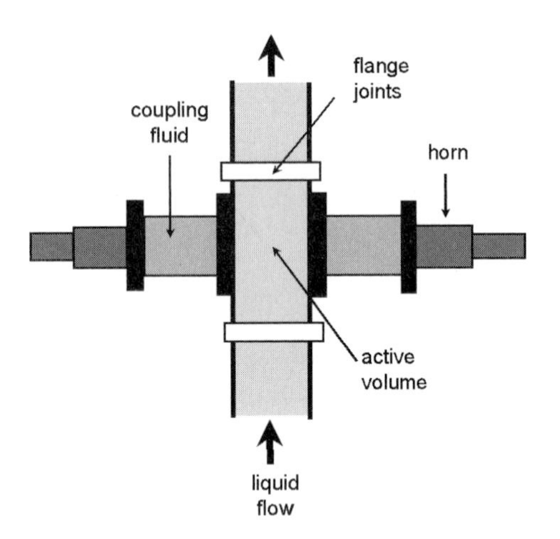

Figure 4.14 Tube reactor with liquid coupling to probe systems

pipe has a tube in its centre carrying cooling liquid (Fig. 4.15). The cooling system occupies the space which would otherwise be the focus of acoustic energy. In acting as both a temperature control and a co-axial reflector of the sound energy this could well prove to be an efficient design for a flow reactor. The annular space between the resonating pipe and the cooling tube can be used to hold solid reagent or catalyst while the reaction mixture is pumped through the system.

There is another method of transmitting ultrasonic vibrations into a tube using a probe system. The horn is directly attached to an annular collar which acts as a cylindrical resonator (Fig. 4.16). The collar has screw fittings to take one or two stainless steel pipes accurately machined so that the remote end is a null point and may be further coupled to other pipework. At an operating

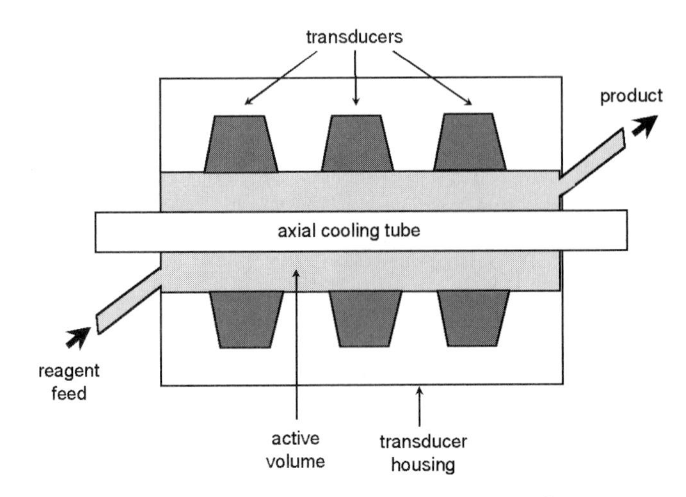

Figure 4.15 Cylindrical reactor with core cooling

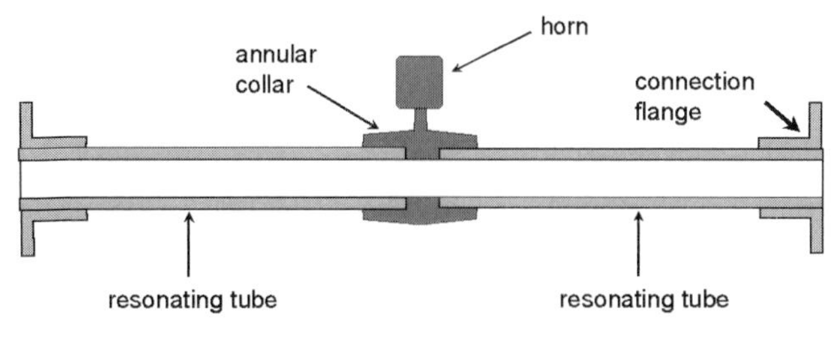

Figure 4.16 Tube reactor driven by central horn

length of 1.2 metres and internal tube diameter of 42 mm the unit can be driven at 2 kW per 1.2 metres with 80 per cent efficiency. The maximum resonance power then operates on the flowing liquid at half-wavelength distances along the tube. An advantage for this system is that when the steel tubing becomes eroded it can be easily replaced.

Resonating bar inserts

In order to introduce ultrasound into a medium flowing through a tube it is only necessary to place a vibrating source (e.g. probe) in contact with that liquid. A neat method of doing this is via the coaxial insertion of a radially emitting bar into the pipe containing the flowing liquid; this would require minimal change to existing pipework. This type of system was originally developed as a form of submersible transducer for cleaning the inside of barrels and tubes immersed in a cleaning fluid.

One system, developed by the Swiss company Telsonic, consists of a hollow, gas filled, tube sealed at one end and driven at the other by a standard piezo transducer (Fig. 4.17). The device looks like a conventional probe system, but is significantly different in that the sealed end is at a point of minimum motion and is not the major source of vibrational energy. Instead the ultrasound is emitted radially at half-wavelength distances along its length. There is the potential to unblock the end and use the system as a flow tube. Designed and marketed for cleaning, there is currently no information on its potential for chemical applications.

A design introduced by the German company Martin Walter involves a cylindrical bar of titanium (cut to a precise number of half-wavelengths at the frequency used). Opposing piezoelectric transducers are attached at each end connected through a central wire (Fig. 4.18). With these transducers operating together in a push–pull mode the 'concertina' effect makes the bar of

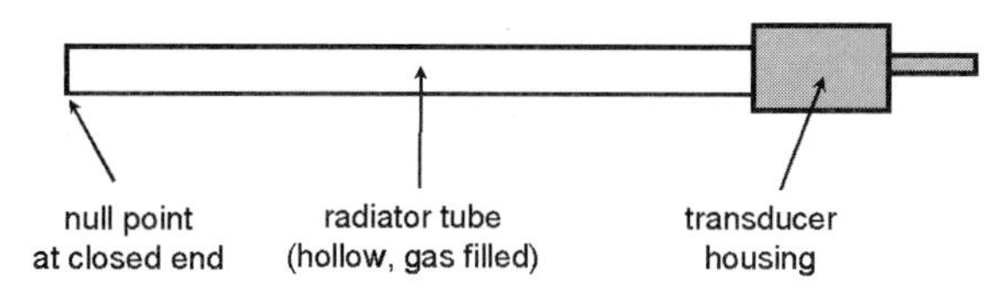

Figure 4.17 Single transducer resonating bar insert

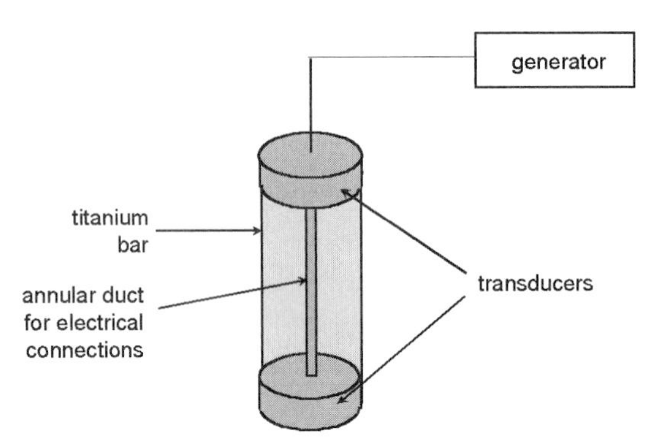

Figure 4.18 Push-pull resonating bar insert

metal expand and contract at half-wavelength distances along the entire length. Any erosion in this system would not affect the resonant length, and since the bar is essentially solid metal, material loss by erosion is not a major problem.

The liquid whistle

The liquid whistle was originally introduced as an efficient mixing/homogenisation system and this still remains one of its main applications. The construction and mode of action of the instrument has been described in Chapter 1. These devices differ markedly from the more usual bath and probe types in that they derive their power from the medium (by mechanical flow across the blade) rather than by the transfer of energy from an external source to the medium. The majority of the chemical effects observed on using whistle type transducers for the sonication of heterogeneous reactions can be attributed to the generation of very fine emulsions rather than the ultrasonic irradiation itself. The use of the liquid whistle type of ultrasonic generators for homogenisation has increased dramatically since the Second World War. It was as far back as 1927 when Wood and Loomis reported that oil and water could be emulsified on sonication in the same beaker using quartz piezoelectric transducers [11]. However, at that stage in the development of ultrasonic equipment it was quite difficult to see how transducers driven by quartz crystals could be employed on an industrial scale, and so ultrasonic homogenisation remained a curiosity until 1948. In that year Janovski and Pohlmann highlighted the economic advantages to be gained from the use of a liquid whistle compared with magnetostrictive transducers of the type available at that time [12]. In 1960 a series of experiments was undertaken to compare four methods then in common usage for the emulsification of mineral oil, groundnut oil, and safflower oil [13]. The results proved that a homogeniser, which operated via a liquid whistle, was superior to three other types of apparatus namely a colloidal mill and two types of sonicator, one of which employed a quartz crystal and the other a barium titanate transducer.

The method has several advantages over flow systems involving probes:

1. The system was specially developed for such processes as emulsification, homogenisation, and dispersion.

2. With no moving parts, other than a pump, the system is rugged and durable.

3. It can be used for the processing of flow systems and, as such, has immediate possibilities for scale-up and can be installed 'on-line'. In this way large volumes can be processed as is the case in the manufacture of such items as fruit juices, tomato ketchup, and mayonnaise.

The intensity of sonication is not as high as that of the probe, however, and this leads to some disadvantages:

1. The ultrasonic power obtained from such a system is limited, so that ultrasonically initiated and driven reactions are less likely to be effected by whistle sonication.

2. The optimum frequency of blade vibration (normally designed to be 20 kHz) relies upon a rapid pumping of the reaction. Clearly this is not possible for viscous materials.

3. The underlying mode of action, involving as it does the flow induced cavitation near a steel blade, immediately suggests that there may be a problem of blade erosion by particulate matter. This does not seem to be a major problem in working systems.

The vibrating tray

This device was originally designed by the Lewis Corporation for the processing of coal or metal ores at rates up to 20 tons per hour (Fig. 4.19) [14]. In the particular case of coal recovery from waste tips, the process involves mixing the coal 'waste' with equal quantities of water and, after crude screening to

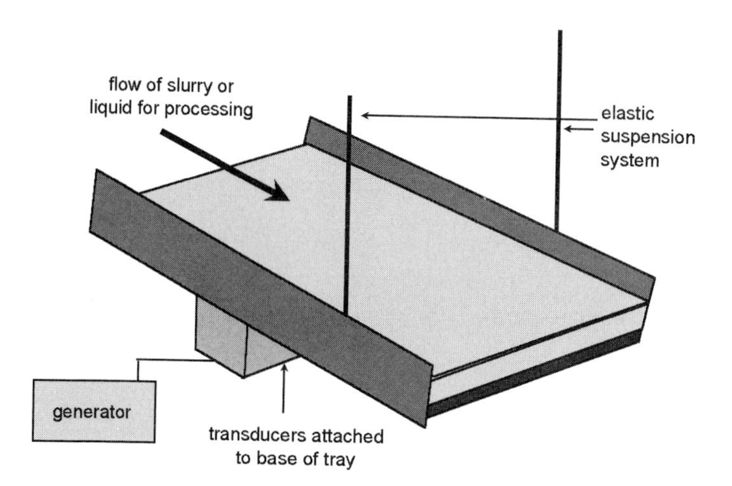

Figure 4.19 Shaking tray system

remove rocks, allowing gravity to carry the mixture down the vibrating tray (20 kHz). With a residence time of only a few seconds this process yields marketable low-ash coal product, with clay and sand suspended in the wash water which can be allowed to settle in a pond from which the water is recycled for further processing.

The vibrating tray is generally used under ambient conditions and is both large scale and extremely robust. Design modification to allow the unit to be installed within a pressure vessel would permit the processing to be used at elevated temperatures with complete containment of vapours. This then becomes a piece of heavy-duty chemical processing equipment for solid/liquid reactions or extractions.

4.4 Low-frequency high-power processing system

A significantly different system has been introduced to large-scale processing by Arc Sonics of Canada, and this involves audible frequency vibrations generated in a large cylindrical steel bar. The bar is driven into a clover leaf type of motion by firing three powerful magnets in sequence which are located at each end of the bar. The bar is supported by air springs, so that the ends and the centre are then caused to rotate at a resonance frequency depending on the bar's size (Fig. 4.20). One such unit, operating at a power of 75 kW, drives a bar which is 4.1 m in length and 34 cm in diameter at its resonance frequency of 100 Hz. The bar itself weighs 3 tonnes and produces a vibrational amplitude of 6 mm. Other units using smaller sized bars operating at higher, though still audible, frequencies have also been built. These vibrating bar systems can be used in chemical processing applications by fixing a robust cylindrical steel cell to each end of the bar. Material in the form of a liquid or slurry can then be pumped through the cells in order to perform operations such as mixing, grinding, and the destruction of hazardous waste. Hard spherical grinding balls are often added to the cells to assist in these processes.

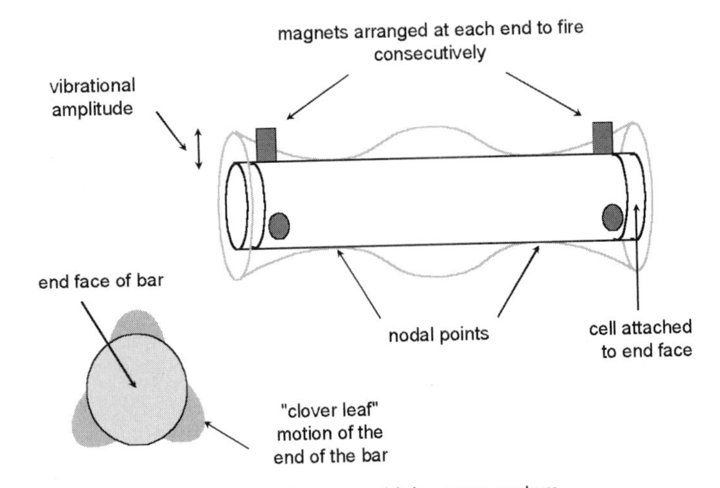

Figure 4.20 Low frequency high power system